The Spectrum Handbook 2013

- Spectrum valuation framework
- Sources and timing of new mobile spectrum
- Break-out of mobile wireless spectrum allocations, demand and fundamental business dynamics
- International comparative analysis of spectrum regulation and allocation
- Explanation of Spectrum and wireless technology issues
- Overview of fundamental regulatory and licensing issues
- Explanation of spectrum usage in all major frequency bands

Acknowledgements

We would like to thank the following people who generously donated their valuable time and expertise providing input on various drafts of this publication:

Juan Deaton (Idaho National Laboratory), Howard Del Monte (formerly with Ofcom), Barlow Keener (Keener Law Group), Timothy Horan (Oppenheimer & Co.), Bruce Jacobs (Pillsbury Winthrop), Jan Kruys (Spectrum Consult), Neal Mellen (Wireless Spectrum Management, LLC), Craig Schultz (Northwestern Univ. Law School), Peter Stanforth (Spectrum Bridge, Inc.), Drew Svor (Sheppard Mullin, LLP), Brian Weimer (Sheppard Mullin, LLP)

Any remaining errors or omissions remain mine.

Praise for *The Spectrum Handbook 2013*

"Reforming the way we manage radio frequency spectrum is essentially a multi-disciplinary endeavor that requires the co-evolution of technology, business models, and policy. Armand Musey's handbook is a boon for academics and practitioners seeking to engage the difficult issues of valuing, transacting, and planning wireless investments because it provides a comprehensive and holistic overview of the technical, legal, and economic challenges of spectrum management in a single and easily accessible volume that still manages to go deep in addressing the challenges confronting the experts today. Bravo, to Armand!"

William Lehr, PhD
Research Scientist, Massachusetts Institute of Technology

"*The Spectrum Handbook 2013* is a fantastic resource for all those who interested in one of the most valuable assets in our nation: wireless spectrum. Anyone who has ever tried to figure out even the most basic issues about spectrum allocation and utilization knows that what should be a relatively straightforward task can be very challenging in light of the complex FCC and other rulings that govern the vast wireless universe. Armand Musey has created an excellent desktop resource for everyone involved in the wireless industry."

Brian D. Weimer
Communications Practice Leader, Sheppard Mullin

"Spectrum is the electromagnetic lifeblood of the digital economy. This handbook provides a thorough analysis of the legal and economic aspects of spectrum management, both from the US and international perspective. It is a valuable reference for wireless entrepreneurs looking for new opportunities to invest."

Thomas Welter
Chief Frequency Officer, SFR (France)

Summit Ridge Group, LLC

Summit Ridge Group is the premier provider of business valuation and financial advisory services in the satellite, media and telecom sectors. Our services include business appraisal, intangible asset valuation, strategic consulting, custom research and litigation support. We provide these services for a range of client needs including transactional support, financial reporting, financial analysis, market analysis, strategic analysis, regulatory support and litigation support/expert witness testimony. Our deep industry focus means that we can quickly grasp a client's issues and provide rapid comprehensive responses and solutions.

Visit us at:

Web: www.SummitRidgeGroup.com
Blog: www.SummitRidgeGroup.com/blog
Twitter: @armandmusey

J. Armand Musey, CFA, JD/MBA

J. Armand Musey is a leading expert in the satellite, media and telecom sectors as well as president and founder of Summit Ridge Group. He is a former top-ranked Wall Street research analyst and an investment banker. Armand has been involved with a wide breadth of companies in the communications sector over the past 15 years. This gives him a deep understanding of the complex web of relationships underlying the sector's competitive dynamics and the associated financial and regulatory issues. Armand speaks at major industry conferences and is regularly quoted in the press as an expert in media and telecom finance.

Mr. Musey holds an MBA from Northwestern University's Kellogg Graduate School of Management and a JD (cum laude) from Northwestern University Law School. He also has an MA from Columbia University and a BA (with honors) from the University of Chicago. He is a member of the New York Bar and Federal Communications Bar Association. Armand is also a Chartered Financial Analyst (CFA) and a member of the New York Society of Securities Analysts where he chaired the Corporate Governance Committee from 2007-2009.

The Spectrum Handbook 2013

J. Armand Musey, CFA
Founder/President

Greta Golz
Research Associate

For additional copies or inquiries please contact:

Summit Ridge Group, LLC
535 Fifth Avenue, 4th Floor
New York, NY 10017
Phone: +1.646.843.9850
Email: Admin@SummitRidgeGroup.com

Summit Ridge Group, LLC
535 Fifth Avenue, 4th Floor
New York, NY 10017
www.SummitRidgeGroup.com

Published in the United States of America in 2013
By Summit Ridge Group, LLC

ISBN: 978-0-9892962-0-5

The Spectrum Handbook 2013

IMPORTANT DISCLOSURES VIII

CRITICAL DATES IN SPECTRUM HISTORY 1

EXECUTIVE SUMMARY 3

PART ONE: BACKGROUND ON SPECTRUM AND ITS REGULATION 5

I. ELECTROMAGNETIC SPECTRUM AND RADIO WAVES 6

A. Electromagnetic Spectrum 6

B. Radio Waves 7

C. Bandwidth and Capacity 8

II. DESCRIPTION OF MAJOR SPECTRUM BANDS 11

A. Very Low Frequency (VLF): Under 30 kHz 12

B. Low Frequency (LF): 30 kHz to 300 kHz 13

C. Medium Frequency (MF): 300 kHz to 3000 kHz 13

D. High Frequency (HF): 3 MHz to 30 MHz 14

E. Very High Frequency (VHF): 30 MHz to 300 MHz 15

F. Ultra High Frequency (UHF): 300 MHz to 3 GHz 15

G. Super High Frequency (SHF): 3 GHz to 30 GHz 16

H. Extremely High Frequency (EHF): 30 GHz to 300 GHz 16

III. OVERVIEW OF REGULATORY PROCESS FOR SPECTRUM ALLOCATION 18

A. Background 18

B. The Role of the International Telecommunications Union 18

C. Other International Coordination Entities 22

D. National Spectrum Bodies 23

E. The Role of the Federal Communications Commission and the National Telecommunication and Information Administration 24

F. Regulatory Challenges are Significant 30

PART TWO: ECONOMIC AND REGULATORY DYNAMICS 39

IV. MAJOR COMMERCIAL AND POLICY SPECTRUM ISSUES 40

A. General Spectrum Dynamics 40

B. Mobile Wireless Industry Spectrum Dynamics 48

C. Unlicensed Spectrum/Spectrum Sharing 53

D. Broadcasters (Television and Radio) 55

E. Satellite Communications Spectrum Dynamics 57

F. Government/Military Spectrum Dynamics 60

V. SOURCES AND TIMING OF NEW MOBILE SPECTRUM 64
A. National Broadband Plan and Television Incentive Auction 64
B. 65 MHz in Required Auctions per Spectrum Act 73
C. Other Sources of Additional Mobile Spectrum Capacity 74
D. Government Spectrum Sharing 77

VI. SPECTRUM VALUATION ANALYSIS 83
A. Income Approach 84
B. Market Approach - Comparable Transactions 85
C. Cost Approach - Replacement Method 86
D. Econometric Modeling 87
E. Other Considerations 88
F. Valuation Challenges 88
G. General Valuation Observations 91
H. Valuation Trends 97

VII. GLOBAL PERSPECTIVES 103
A. General Trends 104
B. ITU Region 1: Europe, Middle East and Africa 106
C. ITU Region 2: The Americas 111
D. ITU Region 3: India and Asia 114

VIII. ECONOMIC AND REGULATORY CONCLUSIONS 117

PART THREE: SPECTRUM BAND DESCRIPTIONS AND ANALYSIS 119

IX. MOBILE WIRELESS SERVICE BAND DESCRIPTIONS 120
A. Upper and Lower 700 MHz bands 120
B. Advanced Mobile Phone Systems (AMPS): 800 MHz 124
C. Personal Communications Service (PCS): 1.8 GHz and 1.9 GHz 126
D. L-Band (1.6 GHz) and S-Band (2.0 GHz) 127
E. Advanced Wireless Spectrum (AWS): 1.7 GHz and 2.1 GHz 127
F. Wireless Communication Service (WCS): 2.3 GHz 130
G. BRS/EBS (2.5 GHz) 132

X. RADIO AND TELEVISION BROADCASTING BAND DESCRIPTIONS 134
A. AM Radio: 540 kHz to 1700 kHz 134
B. FM Radio: 92.1 to 107.9 MHz 135
C. VHF Television: 54 to 88 MHz and 174 to 216 MHz 136
D. UHF Television: 470 to 698 MHz 136

XI. SATELLITE FREQUENCY BAND DESCRIPTIONS 137
A. L-Band: 1.4 GHz and 1.6 GHz 138
B. S-band: 2.0 to 2.7 GHz 139

C. C-band: 3.4 to 6.725 GHz 139
D. Ku-band: 10.7 to 18 GHz 140
E. X-Band: 7 to 11.2 GHz 140
F. Ka-band: 26.5 to 40.0 GHz 140
G. V-band: 50 GHz to 75 GHz 141

XII. GOVERNMENT/MILITARY FREQUENCY USE DESCRIPTIONS 143
A. U.S. Federal Agencies Uses Large Amounts of Spectrum 143
B. Defense Is Largest Government User 143
C. Government Spectrum Trends 144

XIII. VERY HIGH FREQUENCY (VHF) SERVICES 146
A. Lower Paging: 35, 43, 152 & 158/159 and 454 MHz 146
B. Radio Control Service: 72 to 73 MHz and 75.4 to 76 MHz 146
C. VHF Public Coast: 150.8 to 162.025 MHz 147
D. Basic Exchange Telephone Radio Service (BETRS) 148
E. Automated Marine Telecommunication System (AMTS) 149
F. Former Interactive Video and Data Service: 218-219 MHz 149

XIV. ULTRA HIGH FREQUENCY (UHF) SERVICES 151
A. Personal Locator Beacons (PLBs): 406 MHz 151
B. Family Radio Service (FRS)/General Mobile Radio Service (GMRS): 462-467 MHz 151
C. Specialized Mobile Radio: 800 MHz 151
D. Air Ground Service: 800 MHz 152
E. Narrowband Personal Communication Service (PCS) 153
F. Location and Monitoring Service (LMS)/Industrial Scientific and Medical (ISM): 902 to 928 MHz 154
G. Multiple Address Systems: 928/959 MHz and 932/941 MHz 155
H. Upper Paging Band: 929 and 931 MHz 156
I. Mobile Wireless Services 157

XV. SUPER HIGH FREQUENCY AND EXTREMELY HIGH FREQUENCY 158
A. Super High Frequency (SHF): 3 GHz to 30 GHz 158
B. Extremely High Frequencies (EHF): 30 GHz to 300 GHz 161

APPENDICES 163

APPENDIX I: ITU REGIONS 164

APPENDIX II: RADIO SPECTRUM FREQUENCIES 165

APPENDIX III: FCC GEOGRAPHICAL LICENSE AREAS 167

APPENDIX IV: GLOSSARY 170

IMPORTANT DISCLOSURES

- **This document is not a recommendation to buy or sell securities of any type.** It is designed to facilitate an understanding of communications industry topics. Please consult an appropriate professional advisor before making significant business or investment decisions
    - This document expresses summary views and therefore does not include all views of Summit Ridge Group, LLC or its professionals
    - General views expressed in this report may or may not be applicable to a given situation. Adjustments and/or changes may be needed to address the particular circumstances of that situation
- Forward looking statements expressed in this report are not guaranteed in any manner and may ultimately prove to be incorrect
- Data in this report, including but not limited to financial information, company and industry information was derived from a number of sources that we believe to be accurate. However, we have not independently verified the data and therefore cannot assume responsibility for its accuracy. Please verify any data before using it as the basis for an important decision
- Comments about regulatory issues should not be taken as legal conclusions or advice. Consult an appropriately qualified attorney before making important decisions requiring regulatory analysis
- Views expressed in this report are subject to change. Summit Ridge Group, LLC does not assume responsibility for updating its contents. Please contact us for our most current views

Critical Dates in Spectrum History

The history of wireless communications mirrors that of many fields of science. The Ancient Greeks were the first known group to study what has become known today as spectrum. After many centuries of little progress, significant new technological developments occurred during the Enlightenment. Progress has generally accelerated ever since, responding to the increased needs of modern economies.

- 5th Century B.C.
  - Greek Pre-Socratic philosopher Empedocles opines (correctly) that light travels at a finite speed
- 1600s
  - Isaac Newton opines that light is made-up of particles (1672)
  - Christiaan Huygens postulates the wave theory of light (1678), adualistic understanding (waves and particles) that continues today
- 1700s
  - Leonard Euler (1740s) and Benjamin Franklin (late 1700s) support Huygens' theory against scientific community which support Newton's view
- Early 1800s
  - Thomas Young's slit experiment confirms Huygen's wave theory of light (1803)
- Mid 1800s
  - James C. Maxwell formulates classical electromagnetic theory (1861)
  - International Telegraph Union (later to become the International Telecommunication Union) founded in Paris (1865)
- Late 1800s
  - Heinrich Rudolph Hertz proves electromagnetic waves exist (1886)
  - Guglielmo Marconi transmits and receives the first wireless signal
- Early 1900s
  - Reginald Fesseden makes first voice radio transmission (1900)
  - Guglielmo Marconi transmits first transatlantic wireless signal from Ireland to Canada (1901)
- 1910s
  - Titanic tragedy emphasizes the importance of radio communication (1912)
  - During WW I, the U.S. Navy takes control of all radio technology
  - After WW I, the Radio Corporation of America (RCA) established to take over patent control from the government
- 1920s
  - First radio program transmitted on a daily basis (1920)
  - Gulielmo Marconi discovers short wave radio that can reflect against the ionosphere (1920)

- 1930s
  - Frequency-modulated (FM) radio invented by Edwin Armstrong (1933)
  - The Federal Communications Commission (FCC) established (1934)
- 1940s
  - Wartime military needs encourage deployment of radio services
  - Radiotelephony commercialized
- 1950s
  - First Soviet satellite, Sputnik 1, launched, followed by the first U.S. satellite, Explorer 1
- 1960s
  - Telstar I satellite relays transatlantic television signal (1962)
  - Color television begins (1963)
- 1970s
  - LORAN becomes leading navigation system
  - The FCC allocates 40 MHz for cellular service
- 1980s
  - Cellular spectrum at 850 MHz given to local Bell operating companies [B Block] and allocated to others via comparative hearing, lottery and auction [A Block]
  - Bell Breakup (1984)
  - First cell phone services begin (1984)
  - FCC allocated ISM bands for unlicensed use (1985)
  - First generation of GPS satellites completed (1985)
  - First Internet services provided through dial-up connections
  - Internet commercialized and opened up to the public in 1995
  - First commercial hand-held mobile phone released (1G analog)
- 1990s
  - Second generation (2G) cellular technology (digital) launched
  - First text messages sent from cell phone to cell phone
  - Internet access started switching from dial-up to broadband
  - U.S. Army begins aggressive work on software radios that could dynamically change frequencies (1994)
  - FCC conducts its first spectrum auction (1994)
  - FCC finally publishes Wi-Fi device standards for ISM bands that it originally allocated in 1985 (1997)
  - Digital television broadcasting begins (late 1990s)
- 2001
  - Third generation (3G) cellular technology launched
- 2006
  - Forth generation (4G) cellular technology released in South Korea
    - 4G not launched in the U.S. until 2008 by Sprint Nextel
- 2008
  - First reallocations of U.S. spectrum bands in the 700 MHz range
  - Analog services phase-out to digital service begins
  - Licensing TV white spaces accelerates share spectrum movement

Executive Summary

This Handbook has three objectives: 1) to serve as a primer for explaining the complex issues around the use of electromagnetic spectrum; 2) to analyze, from both an economic and a legal perspective, the regulatory processes being considered or underway to reallocate or change the use of spectrum bands; and 3) to be a reference source for industry professionals. Part I of the Handbook provides an overview of the spectrum and the regulatory process. Part II of the Handbook explains the various available spectrum bands, discussing their range, location, and physical properties and how these impact their ability to be used. An analysis of the current allocation of these spectrum bands in the United States follows. Part III contains detailed explanations of the various spectrum band plans. Throughout the Handbook, we provide links in the footnotes to sources for additional information.

From a macro-perspective, regulators worldwide are currently reallocating spectrum from underutilized applications to the burgeoning mobile wireless broadband applications. Given the needs and importance of wireless broadband, from an economic and social perspective, this trend is likely unstoppable. The FCC is allocating both licensed spectrum (including the broadcast incentive auction) and unlicensed spectrum (including the 3.5 GHz and 5 GHz processes). Unlicensed (shared) spectrum is one approach to minimize disruption from these reallocation efforts and expand utilization is the small but significant spectrum sharing movement. Spectrum sharing is simultaneously proposing to improve spectral efficiency and calling into question the need for licenses altogether.

As a result of the existing processes underway to improve the efficiency of spectrum allocation along with new technologies that further improve efficiency, the extent of "spectrum crunch" (i.e. the apparent lack of available spectrum) is poorly understood and hotly debated. Summit Ridge Group does not believe it is likely to bring information access to a grinding halt in the United States. Rather, we may see temporary congestion while regulators approve new reallocations of spectrum and spectrum-sharing plans, and service providers build out services on new spectrum. These processes, combined with wireless carriers' improved ability to regulate customer data usage—primarily by charging higher fees and/or capping usage, and offloading traffic via Wi-Fi and other technologies—should allow operators to continue to provide reliable service in the face of increasing demand. These trends are also likely to temper the increase in spectrum prices in the future.

PART ONE: BACKGROUND ON SPECTRUM AND ITS REGULATION

I. Electromagnetic Spectrum and Radio Waves

A. Electromagnetic Spectrum

Radio waves occupy a relatively small part of the electro-magnetic spectrum

Electromagnetic spectrum is the range of all frequencies of electromagnetic radiation. Electromagnetic radiation is a form of energy that moves in a wave-like form as it travels through space. These waves have different transmission characteristics depending on their wavelength. Spectrum is often described in terms of its frequency, which is the number of waves that pass a given point per second. The number of waves made in a second is typically referred to in units called Hertz[1] (Hz), where one wave per second is 1 Hz.[2] Frequency may also be referred to by the absolute length (in metric units) of its waves. As electromagnetic waves move at a constant rate, the speed of light, the wavelength and the frequency are inversely related. The longer the wavelength, the lower the frequency and visa versa. Mathematically:

$$frequency\ (in\ Hz) = \frac{299{,}792{,}458\ m/s\ (speed\ of\ light)}{wavelength\ (in\ meters)}$$

and graphically illustrated in Exhibit 1-1 below.

[1] Named after the German physicist Heinrich Hertz (1857-1894) who proved the existence of electromagnetic waves.

[2] Spectrum is commonly referred to in units of kilohertz (kHz) – one thousand waves [hertz] per second, megahertz (MHz) – one million waves [hertz] per second and gigahertz (GHz) – one billion waves [hertz] per second.

Exhibit 1-1: Relationship Between Wavelength and Frequency

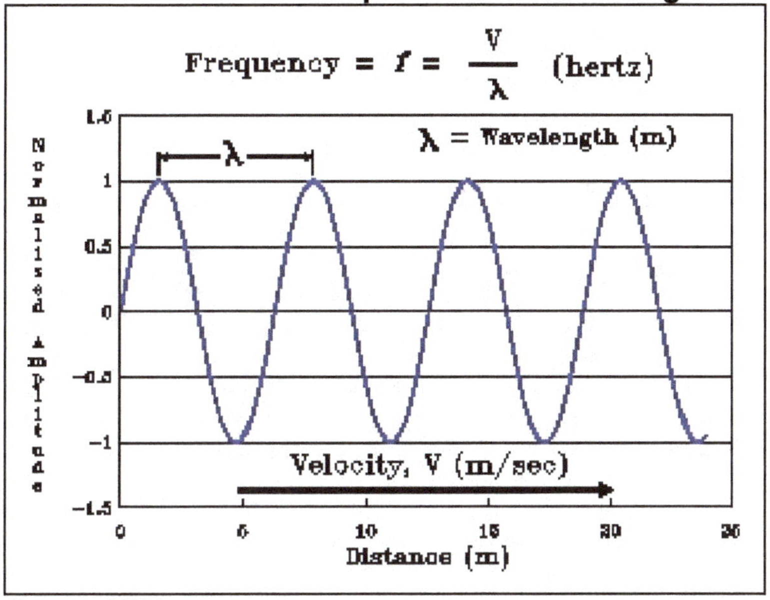

Source: NTIA.

Radio waves make up the part of the spectrum

Radio waves occupy only a relatively small part of the electromagnetic spectrum. Other parts of the electromagnetic spectrum include X-rays, visible light and gamma rays. A brief summary of the radio spectrum is given in Exhibit 2 below. A more detailed analysis is contained in Appendix II at the end of this Handbook.

B. Radio Waves

Radio waves occupy a small portion of the electromagnetic spectrum, namely that between 3 kilohertz (kHz) and 300 Gigahertz (GHz).[3] In the United States, only radio frequencies between 9 kHz and 275 GHz have been allocated for any use, government or commercial, so far.[4] Common allocations are listed in Exhibit 1-2 below. In general, services that require the transmission of signals over long distances, such as navigational services use waves with

---

[3] The U.S. regulatory bodies (the FCC and the NTIA) as well as the International Telecommunications Union define radio spectrum as including all frequencies below 3,000 GHz.

[4] Federal Communications Commission. FCC Encyclopedia. "Radio Spectrum Allocation." Accessed May14, 2013. http://www.fcc.gov/encyclopedia/radio-spectrum-allocation.

low frequencies and thus long wavelengths, while services that are more restricted in terms of the geographical area in which they are provided, such as Wi-Fi devices, use waves with high frequencies and therefore short wavelengths.

Exhibit 1-2: Uses of the Radio Spectrum

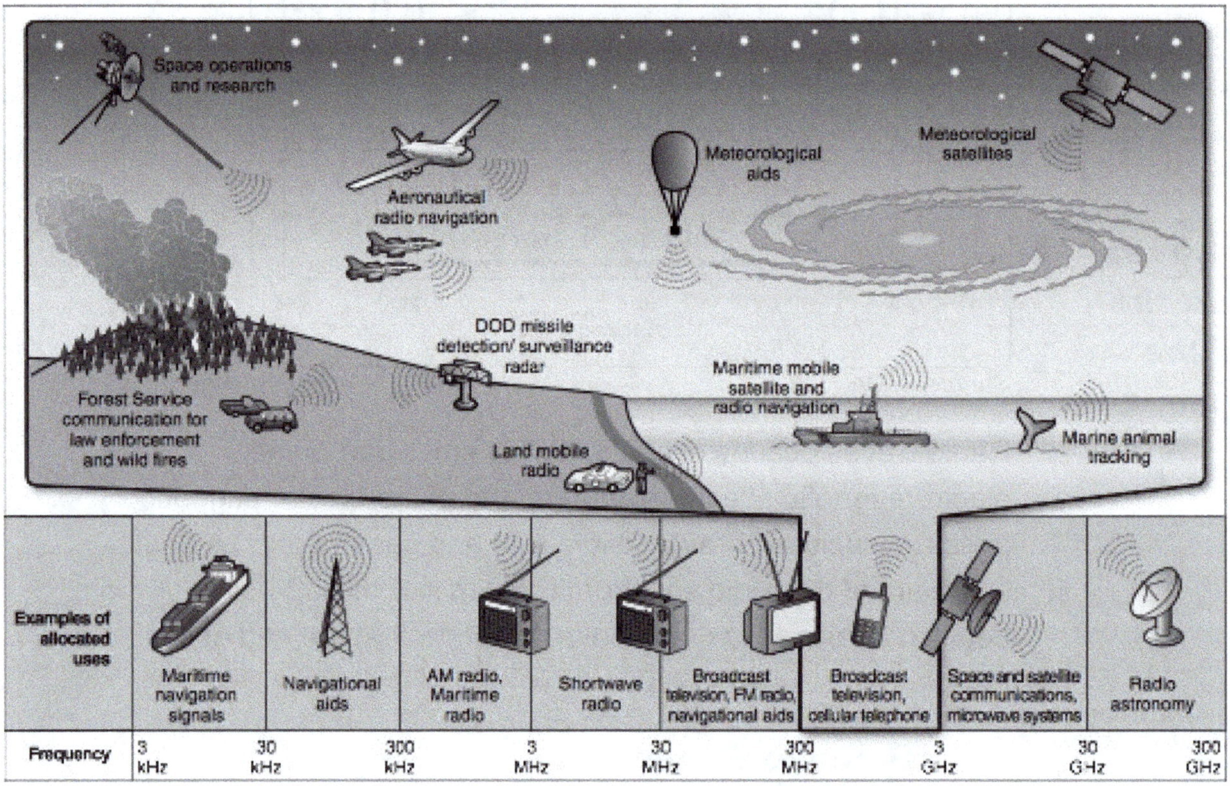

Source: GAO. (2011). Spectrum Management: NTIA Planning and Processes Need Strengthening to Promote the Efficient Use of Spectrum by Federal Agencies.

C. Bandwidth and Capacity

The data capacity of spectrum is generally proportional to the amount of bandwidth used. Generally digital satellite transmission (with its weaker signals) can achieve approximately 1.2 - 1.5 bits/second per hertz of spectrum while newer high-end Long Term Evolution ("LTE") systems can achieve closer to 1.4 bytes/sec per hertz (1 byte = 8 bits). Mathematically, the theoretical limit of bits/sec per hertz is represented by the Shannon-Hartley theorem[5] (often referred to as Shannon's Law). Mathematically:

*Capacity = (Bandwidth in Hertz) * Log_2(1 + Signal/Noise)*

Spectrum capacity can often be increased by reusing it

To date, communication systems have been unable to approach this theoretical limit. However, some advanced systems can achieve utilization at approximately half of this level. This suggests that while there is potential for additional improvement in spectrum technologies, it may be limited.

The demands for data vary by application. Data demands of common applications are listed in Exhibit 1-3 below:

Exhibit 1-3: Data rate requirements of common applications

Application	Data rate
Voice call (low quality)	3 kilobits/second
Voice call (medium quality)	30 kilobits/second
Voice call (high quality)	90 kilobits/second
MP3 music (standard quality)	128 kilobits/second
MP3 music (high quality)	256 kilobits/second
Television (low quality)	1 megabits/second
Television (standard definition)	2 megabits/second
Television (1080i high definition)	5 megabits/second
Telemedicine	10 megabits+/second

Source: Summit Ridge Group, LLC analysis.

Spectrum capacity can often be increased reusing it. Depending on the power levels, topography, type of spectrum, etc., service providers can limit signals to a small area. They can then install another antenna in the adjacent area to allow reuse of spectrum without interferences. This process is known as "cellularizing" and is illustrated in Exhibit 1-4 below:

---

[5] Named after American mathematician, engineer and cryptographer, Claude Shannon (1916 – 2001) who largely invented the idea of digital communication, and American electronics researcher Ralph Hartley (1888 – 1970) who contributed to the foundations of information theory.

Exhibit 1-4: Cellular Reuse

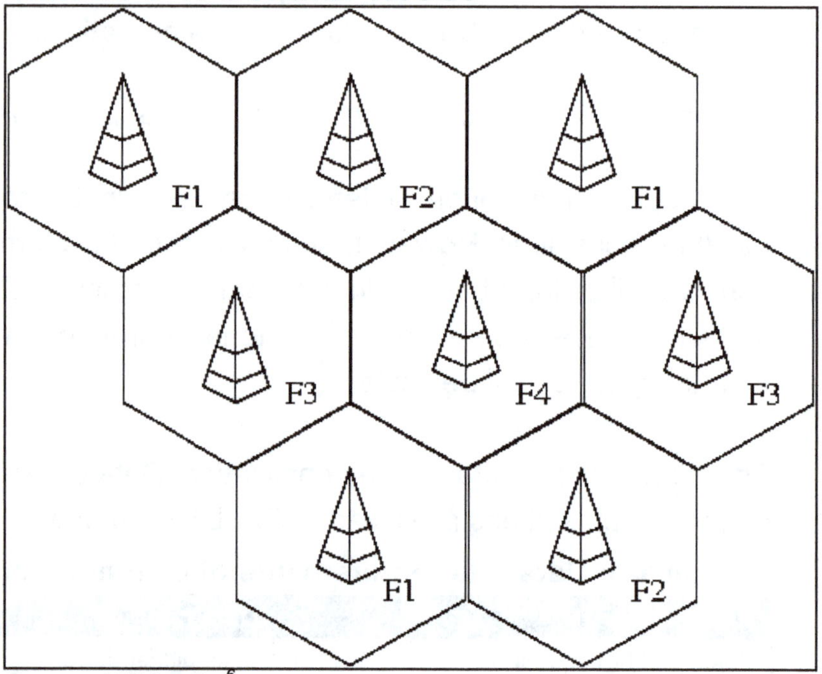

Source: Wikipedia.[6]

Exhibit 1-4 above shows an example of frequency reuse with four frequencies. The image shows an idealized situation with perfectly hexagonal cells. The cell on the top left uses frequency F1. The cells next to it use frequencies F2 and F3. Beyond those cells, another cell uses frequency F1 again. This pattern, with the same frequency never being reused by adjacent cells, repeats across the coverage area. This frequency reuse pattern is typical for a digital wireless system.

---

[6] This file is licensed under the Creative Commons Attribution-Share Alike 3.0 Unported license. User: Mozzarati.

II. Description of Major Spectrum Bands

Lower frequencies have greater distance propagation qualities and better building penetration but require larger antennas and have lower capacity for data transmission

Radio spectrum consists of a range of frequencies that can be used to transmit information both by analog and digital methods. This spectrum ranges from approximately 3 kHz to 300 GHz, although the commercially viable spectrum bands range from about 500 kHz (AM Radio) to approximately 30 GHz (Ka-band satellite). The frequencies have different characteristics with respect to the distance they travel, building penetration (indoor reception), antenna sizes needed and other factors. As a result, some spectrum bands are more suitable than others for certain applications. Therefore the radio spectrum is split into several frequency bands that are assigned particular uses according to the characteristics of the frequencies. In general, lower frequencies have greater distance propagation and better building penetration but require larger antennas and have lower capacity for data transmission.

Most commonly used bands for television, cell phones, pagers, etc. are between 50 MHz and 3 GHz. Devices requiring limited ranges and small antennas, such as Wi-Fi broadband and Bluetooth, tend to use higher frequencies (2.4 GHz and higher) whereas for long distance communication with underwater submarines, the military uses very low frequencies. When radio technology was starting, the early radios used much lower frequencies than modern radios typically use today. However, technology innovations are increasingly making higher frequencies viable for more common use. This is the historical basis for the somewhat confusing naming convention whereby the High Frequency band is actually at the lower end of the currently usable spectrum range and the middle of the usable range is called the Ultra High Frequency band. A full description of all radio bands is beyond the scope of this paper, but the FCC's 168 page description can be found at: http://transition.fcc.gov/oet/spectrum/table/fcctable.pdf. A concise version is available in Appendix II of this Handbook. A summary of major bands is shown in Exhibits 2-1 and 2-2 below:

Exhibit 2-1 Frequency Band Designations (FCC, NTIA & ITU)

ITU Band	Frequency	Band Designation	Abbreviation
4	3 – 30 kHz	Very Low Frequency	VLF
5	30 – 30 kHz	Low Frequency	LF
6	300 – 3,000 kHz	Medium Frequency	MF
7	3 – 30 MHz	High Frequency	HF
8	30 – 300 MHz	Very High Frequency	VHF
9	300 – 3,000 MHz	Ultra High Frequency	UHF
10	3 – 30 GHz	Super High Frequency	SHF
11	30 – 300 GHz	Extremely High Frequency	EHF
12	300 – 3000 GHz	Tremendously High Frequency	THF

Source: Wikipedia, Summit Ridge Group, LLC analysis.

The Institute for Electrical and Electronic Engineers (IEEE) has promulgated a separate frequency band designations using letters.

Exhibit 2-2: IEEE Band Designations

Designation	Frequency
HF	3 – 30 MHz
VHF	30 – 300 MHz
UHF	300 – 1,000 MHz
L-band	1 – 2 GHz
S-band	2 – 4 GHz
C-band	4 – 8 GHz
X-band	8 - 12 GHz
K_u-band	12 – 18 GHz
K-band	18 – 27 GHz
K_a-band	27 – 40 GHz
V-band	40 – 75 GHz
W-band	75 – 110 GHz
G-band	110 – 300 GHz

Source: www.radioing.com and Summit Ridge Group, LLC analysis.

A. Very Low Frequency (VLF): Under 30 kHz

Due to their long wavelengths (over 10 kilometers), transmissions in this spectrum can "skip" over mountains and generally follow the curvature of the earth. In the early part of the twentieth century this VLF was used for sending Morse code across the Atlantic. Current commercial uses of this spectrum include clock radios, heart monitors (at very low power levels), and military communication to submarines near the surface (at much higher power levels) and various experimental uses.

B. Low Frequency (LF): 30 kHz to 300 kHz

LF transmissions also follow the curvature of the earth, bounce off the ionosphere and even deflect off mountains. They can also penetrate the ocean to approximately 200 meters. This spectrum is used for time signals, RFID tags, some experimental purposes, submarine communications, aircraft navigation and some AM radio transmission in parts of Europe, Asia and Africa.

Position location services in the U.S. were formerly provided near the 100 Hz frequency by the U.S. government using a system named LORAN...

Position location services in the U.S. were formerly provided near the 100 kHz frequency by the U.S. government using a system named LORAN (the former Soviet government used an analogous system called CHAYKA). In recent decades, users left LORAN and CHAYKA in favor of Global Positioning Satellites (GPS). The U.S. Coast Guard and the Russian government terminated transmission on their systems and shuttered Loran operations in 2010. The Loran-type system is still used to enhance GPS signals in Europe using a transmission systems known as EUROFIX. A similar system is used in Saudi Arabia under the name Saudi Positioning System (SPS). There is some industry discussion in the United States of integrating an enhanced LORAN, called eLORAN as back-up to GPS for critical applications.

C. Medium Frequency (MF): 300 kHz to 3000 kHz

MF transmissions propagate both in the sky and across damp earth for up to hundreds of miles. Skyward signals can bounce off the ionosphere extending their distance to even thousands of miles. The ability of signals to bounce off the ionosphere varies by season and by time of day, generally increasing significantly at night.

This band includes allocations for navigational beacons (in the 190 to 435 kHz range) and AM radio (535 to 1705 kHz in the U.S. and 526.5 to 1606.5 kHz in Europe). There are many ship to shore radio communications allocations between 1600 and 2850 kHz. The international distress frequency for maritime voice is at 2182 kHz and assigned to channel 16 on the Marine VHF band.

D. High Frequency (HF): 3 MHz to 30 MHz

Despite the name "High Frequency," the frequency is actually quite low, but capable of transmitting signals long distances

Despite the name "High Frequency," the frequency is actually quite low for broadcasting, but capable of transmitting signals long distances. More specifically, by reflecting off the ionosphere, signals in this spectrum range can travel across continents. The primary use of this frequency is for short-wave radio services between 5.950 MHz and 26.100 MHz. These are broadcast services designed to be received by the general public in foreign countries.

High Frequency signal propagation is impacted by atmospheric conditions. As atmospheric conditions change significantly with the seasons, the ITU holds a High Frequency Coordination Conference twice a year to coordinate shortwave broadcasting schedules. The changing propagation of the frequency has made it a favorite with amateur radio operators who often experiment to test transmission distances at various times. Spectrum in the High Frequency range is also used for emergency ship to shore communication and aviation communication.

1. Citizen's Band Radio

The Citizen's Band (CB Radio) is allocated the frequency between 26.965 and 27.045 MHz in the United States. The CB Radio Band is divided into 40 channels of 10 kHz each, which can be used for unlicensed voice communication with approved devices. They are typically used for short-range communication and use is regulated under Title 47, Part 95 of the Code of Federal Regulations (the "CFR").

2. Extensive Government Use

Approximately two-thirds of the high frequency spectrum is controlled by various government agencies. Recent advancements, particularly automated link establishment (ALE) technology, have made high frequency communication more reliable during changing atmospheric conditions, leading to growth in the use of the frequency by government users. The National Communication Systems (NCS) runs a program called the Shared Resource High Frequency Radio Program (SHARES) that facilitates cooperation between government High Frequency users during emergencies. Members agree to relay radio messages from any member who is experiencing an emergency to the intended recipient.

Unlike the High Frequency Spectrum, VHF signals do not bounce off the ionosphere

3. Maritime Use Slowing

As satellite communication has become less expensive and more reliable, maritime use of HF communication has declined. Several providers, including AT&T, have exited the business. Some of the remaining maritime HF operators have received waivers for secondary terrestrial uses and are seeking to build businesses based on those licenses.

E. Very High Frequency (VHF): 30 MHz to 300 MHz

The VHF spectrum includes a number of services. Frequencies in this spectrum work well for short distances – somewhat longer than range of sight. Unlike the High Frequency spectrum discussed above, VHF signals do not reflect off the ionosphere preventing them from extending their distance via ionospheric "bouncing" like HF transmissions. Common VHF applications include ship-to-shore radio, paging and certain radio-controlled devices.

See Section XIII for a detailed description of VHF services.

F. Ultra High Frequency (UHF): 300 MHz to 3 GHz

Spectrum in the UHF band is the most versatile for many modern communications services. Frequencies in this UHF band range from ten centimeters to one meter in length. UHF radio waves do not bounce off the ionosphere. Hills and large buildings generally block UHF signals but they can penetrate buildings, particularly at the lower end of the range. Due to their relatively short wavelength, consumer equipment on this frequency band can use small antennas with significant data-carrying capacity. Services in this band include locator beacons, air ground service, most mobile telecom services (covered in Section IX of this Handbook), television (see Section X) and Wi-Fi service.

See Section XIV for a detailed description of UHF services.

G. Super High Frequency (SHF): 3 GHz to 30 GHz

Super High Frequency is used primarily by satellite-based systems

Spectrum in this range has superior capacity to carry data on a bit per MHz basis. But propagation is largely limited to line of sight reception. Super High Frequency is used primarily by satellite-based systems (described in detail in Section XV). Exceptions include the 4.6 GHz General Wireless Communication Service (GWCS), the 24 GHz band and LMDS (Local Multipoint Distribution Service). In addition to its original primary band at 2.4 GHz, Wi-Fi operates in the 5 GHz band. The FCC recently announced plans to expand Wi-Fi spectrum to include parts of the 5.9 GHz band.

See Section XV for a detailed description of satellite services, most of which are offered in the SHF band.

H. Extremely High Frequency (EHF): 30 GHz to 300 GHz

Spectrum at this level and above is sparsely utilized. It is subject to severe rain fade and has poor ability to penetrate buildings and foliage. Nevertheless, portions of the lower end of the range including the Q-band (33 to 50 GHz) and the V-band (50 to 75 GHz), which are allocated for satellite and short distance terrestrial broadband may be increasingly used in the future. In particular, the

39 GHz band is attracting industry attention for terrestrial broadband delivery, while the V-band is used in inter-satellite links between the U.S. Milstar 1 and 2 military satellites. Part of the V-band (57 to 64 GHz) is allocated for unlicensed use and the upcoming Wi-Fi Standard IEEE 802.11 ad will operate at 60 GHz and allow for transfer rates of up to 7 Gb/sec.

III. Overview of Regulatory Process for Spectrum Allocation

Any given frequency band generally cannot be used at the same time by different users or applications without interference. And some frequencies are superior to others for certain applications. This creates the need for the enormous regulatory project of efficiently managing spectrum to maximize its value to society. Regulators manage spectrum through a two-part process of 1) allocating spectrum bands for the most appropriate applications; and 2) coordinating use between users on a given spectrum band.

A. Background

The first major goal of spectrum regulation is to allocate spectrum to the most technically appropriate applications in proportion to the social need for those applications. This process is very much akin to zoning in a real estate context. If a town decides that it needs more industry, for example, it may expand industrial zoning in areas of the town most suited to industry. Likewise, if more mobile broadband is needed, regulators may allocate additional spectrum for mobile broadband in the most suitable spectrum zones. As with real estate zoning, spectrum regulations may include restrictions to prevent interference with neighboring users.

[I]ntergovernmental treaties proposed by the ITU need governmental approval by the affected countries

The second major goal of spectrum regulation is to coordinate its use among users. Frequently regulators allocate use by issuing licenses to use portions of spectrum for particular purposes with specific parameters. Alternatively, they may allow spectrum to be used on an unlicensed basis, subject to certain priority rules. Regulators also maintain processes for resolving disputes between licensees. Spectrum is allocated and managed through a series of international and national regulatory bodies.

B. The Role of the International

Telecommunications Union

The ITU promotes global cooperation in the use of spectrum

Based in Geneva, Switzerland, the International Telecommunications Union (ITU) is an agency associated with the United Nations. It specializes in promoting cooperation in the use of international information and communication technologies, including the allocation, standardization and regulation of global radio spectrum. The ITU was founded in Paris in 1865, as the International Telegraph Union. It took its present name in 1934. Currently, the ITU has 193 member nations and 700 private sector entities and academic institutions that have joined the Union as non-voting Sector Members. The ITU rules, including spectrum allocations are published in its annual report *Radio Regulations.*[7] Individual countries may deviate from ITU rules, including spectrum allocations, so long as those deviations do not interfere with other countries that comply with ITU rules.

The ITU divides the world into three regions, Regions 1 to 3. Europe and Africa are in Region 1, the Americas are in Region 2, while Asia is in Region 3

Despite its active involvement in international spectrum coordination, the ITU, as in many areas of international law, does not have an effective enforcement mechanism. It largely depends on its member countries to abide by its rules out of their long-term interest in having a coherent international spectrum policy. Entities using spectrum are thus subject to the rules of the national government(s) who issued their license(s), not the ITU. However, many global entities lobby the ITU because ITU rules are likely to be adopted by most counties.

The ITU divides the world into three regions, Regions 1 to 3. Europe and Africa are in Region 1, the Americas are in Region 2, while Asia is in Region 3. See Appendix I of this Handbook for a map of the ITU Regions.

The ITU also hosts conferences, most recently the World Conference on International Telecommunications 2012 (WCIT-12), where it proposes intergovernmental treaties addressing telecommunication issues such as the global allocation of spectrum

---

[7] Available at: http://www.itu.int/pub/R-REG-RR-2012.

and international tariffs. Once adopted, these treaties must pass the national approval processes of the affected countries, although the ITU itself is independent.

The ITU rule making process is time-consuming. To affect a change in an ITU policy or standard, one must add the item to the docket a year or so in advance. Frequently no decision is made on an item, which is then carried over to the next conference. Thus ITU policy change is typically measured in years.

ITU rule making process is time-consuming

The ITU has three major divisions. The Radio Communications Sector (ITU-R) coordinates radio frequency spectrum and satellite orbits. The Standardization Sector (ITU-T) produces and revises technical standards for a variety of communications technologies from IPTV to network communication protocols. The Development Sector (ITU-D) promotes initiatives to expand communications access in underserved areas.

1. The Radio Communications Sector (ITU-R)

For spectrum issues, the ITU-R is the most important division of the ITU. Its specific role within the ITU framework is to:

1. Affect the allocation of spectrum bands, the allotment of radio frequencies, and the registration of radio frequency assignments of any associated orbital position in the geostationary satellite orbit to avoid interference between radio stations of different countries;
2. Coordinate efforts to eliminate interference between radio stations of different countries and to improve the use of radio frequencies and the geostationary-satellite orbit for radio communication services.[8]

ITU-R executes its mission by managing five major activities including: 1) World Radio Communication conferences (WRC) that it holds every three or four years. The latest was held in 2012 and the next is scheduled for 2015. The WRC reviews and revises the Radio Regulations - the international treaty governing the use of the

[8] International Telecommunications Union (ITU)

radio spectrum and satellite orbits. Revisions are based on an agenda created by the ITU Council, which is set more than a year in advance. The ITU Council in turn, considers recommendations made by previous WRC conferences. 2) Radio Communication Assemblies (RA) manage the structure, the program and the approval of various radio-communication studies. 3) The Radio Regulations Board (RRB) consists of twelve people who are elected at the Plenipotentiary Conference that meets up to four times a year. The RRB approves rules of procedure and addresses issues that the Radio Regulations cannot resolve and advises on a variety of other issues. 4) Study Groups: Over 1,500 specialists from telecommunication organizations and administrations throughout the world participate in the study groups focused on compiling handbooks, drafting technical bases for Radio Communication Conferences, and developing draft recommendations. 5) The Radio Communication Advisory Group (RAG) reviews priorities and strategies of the ITU-R, monitors progress of the study groups, supervises the Study Groups and facilitates cooperation and coordination with other organizations and with the other ITU Sectors.

In recent years, the ITU-R has placed a great deal of emphasis on International Mobile Telecom

2. ITU-R Priorities

The proliferation and increasing importance of communications devices using different frequencies have increased the importance of spectrum coordination. Consumers expect devices such as mobile phones, but also those using Wi-Fi and Bluetooth technologies to work internationally. And hardware manufacturers need spectrum uniformity to keep production costs low. In recent years, the ITU-R has placed a great deal of emphasis on International Mobile Telecom. It has worked to standardize the 800 MHz band as well as the "digital dividend" spectrum in the 700 MHz range. This digital dividend spectrum was spectrum that was made available when television broadcasting migrated from analog signals to more spectrally efficient digital signals.

The ITU-R also takes the lead in identifying spectrum band globally that can be used for mobile broadband. It believes such identification can facilitate international harmonization and thus lower-cost devices and global availability.[9]

C. Other International Coordination Entities

While the ITU sets down basic principles for spectrum use worldwide, geographically-focused entities handle much of the coordination and details. These entities include:

1. Regional Coordinating Bodies

Regional coordinating bodies include the European Conference of Postal and Telecommunications Administration (CEPT) in Europe and the Inter-American Telecommunication Commission (CITEL) for the Americas. These regional coordinating bodies, as the name suggests, coordinate frequency issues related to a given region. They often focus their efforts on cross-border interference issues and frequency harmonization to allow hardware to travel from one country to another.

In addition to official governmental bodies, private standards organizations operate to influence technical standards

2. International Intra-governmental Agreements

In addition to the ITU, regional bodies and national bodies, individual countries frequently have agreements among themselves for the management of various spectrum issues. The FCC, for example, lists over 50 spectrum-related agreements between the United States and Canada alone.[10]

3. Private Groups

In addition to official governmental bodies, private standards organizations operate to influence technical standards for equipment. These groups can be national or international in focus. In the United States, ATIS, a standards organization, focuses on the communications industry. It is accredited by the American National Institute of Standards (ANSI) and is a member of several international bodies.

---

[9] Roberto Ercole, "The Future of Mobile." Accessed May __, 2013. *https://itunews.itu.int/en/2067-The-future-of-mobile.note.aspx.*

[10] For a list of U.S. frequency agreements with Canada, see: http://transition.fcc.gov/ib/sand/agree/can_frequency.html.

International umbrella groups, such as the Third Generation Partnership Project (3GPP) have also formed to create greater global influence. 3GPP has four technical specification groups that coordinate with standards development agencies to set international standards in the wireless mobile industry. Member agencies of 3GPP include ARBIB (Japan), CCSA (China), TTA (Korea), ATIS (USA), ETSI (Europe) and TTC (Japan).

D. National Spectrum Bodies

Virtually every country has a national coordinating body to manage spectrum within its borders

National spectrum bodies have two primary roles. The first is to administer spectrum policies in their countries and the second is to issue spectrum licenses to users to operate under those policies. As mentioned earlier, countries may deviate from ITU standards so long as the deviation does not cause interference with countries who follow ITU rules. Virtually every country[11] has a national coordinating body to manage spectrum within its borders. In many cases these are connected with the national postal service, due to historical reasons and the common relationship both have to communications. These agencies include the Federal Communications Commission (FCC) in the United States, Ofcom in the United Kingdom, Bundesnetzagentur (BNetzA) in Germany, and the Ministry of Communications and Information in South Korea. Spectrum users must obtain the appropriate permissions/licenses from the national regulatory agencies in the countries in which they operate.

National spectrum bodies also dictate terms for spectrum licenses including length, method of acquisition (fee, auction, assignment etc.), qualifications of licensees and other obligations of licensees. There are three general types of spectrum regulation models: 1) Command and control whereby the government allocates and revokes spectrum usage rights by unilateral decision; 2) Property rights or quasi property rights whereby the government issues a

[11] The few exceptions include the Vatican, Somalia and a handful of other countries in exceptional circumstances.

transferable license to exclusively use spectrum; and 3) The Commons model whereby spectrum is open for use without a licenses. These are shown in Exhibit 3-1 below:

Exhibit 3-1: Spectrum Allocation Models

Model	Users	Uses
Command and Control	• Government Agencies • Maritime industry • Aviation industry	• Military • Aeronautical • Maritime Safety • Emergency Services
Property Rights or Quasi Property Rights	• Commercial entities including: broadcasters, wireless service providers and satellite operators	• Mobile voice and data services • Satellite broadcasting • Fixed wireless
Commons	• Individuals • Companies • Government Agencies	• Wi-Fi • Garage door openers • Bluetooth

Source: ITU[12] and Summit Ridge Group, LLC analysis

Most countries now allocate the most valuable spectrum along a property rights model

Most countries now allocate the most valuable spectrum along a property rights model (aside from emergency services and certain government/military uses which few countries would want sold-off and continue to allocate on a command and control model). However, as discussed in Section IV.A.2 there is growing movement towards the commons approach to maximize spectrum utilization. In addition to considering physical characteristics of spectrum and the potential uses, most national regulatory bodies also consider international implications when allocating spectrum. The most significant of these is spectrum harmonization – standardization of spectrum band usage. International harmonization facilitates international mobile phone roaming, satellite interference avoidance, access standardized equipment and addresses other issues.

E. The Role of the Federal Communications

[12] International Telecommunications Union, Broadband Series, "Exploring the Value and Economic Valuation of Spectrum" p.3. Available at: http://www.itu.int/ITU-D/treg/broadband/ITU-BB-Reports_SpectrumValue.pdf

Commission and the National Telecommunication and Information Administration

1. Background

The United States, unlike most countries, maintains two separate agencies to regulate spectrum. First, the Federal Communications Commission (FCC) is an independent agency of the U.S. government that oversees the portion of spectrum in the public domain for private and commercial users. Additionally, the National Telecommunications and Information Administration (NTIA) is the agency within the Commerce Department that regulates and administers the spectrum allocated to government use. The *Federal Register*[13] is the legal source for U.S. frequency allocation for both the FCC and the NTIA.

The FCC ...oversees the portion of spectrum in the public domain for private and commercial users

The FCC was created by the Communications Act of 1934 to manage the allocation of all commercial spectrum available in the U.S. for the public good. The rules governing the FCC are in Title 47 of the CFR. Part 2 of Title 47 of the CFR contains most rules related to radio spectrum allocation. The FCC has allocated spectrum from 9 kHz to 275 GHz to various commercial uses.

The NTIA manages the portion of the spectrum allocated for U.S. government use

Section 305 of the Communications Act of 1934 gives the President the authority to assign all frequencies used by the federal government and in 1978 this power was embedded in the newly created NTIA. Government agencies pay a small below-market fee, to the NTIA for a spectrum assignment. NTIA rules can be found within the *Manual of Regulations & Procedures for Federal Radio Frequency Management,*[14] often referred to as the "NTIA Manual" or the "Redbook." An illustration of the relationship between the FCC and the NTIA is shown in Exhibit 3-2 below.

---

[13] *The Federal Register* can be found at: *http://www.gpo.gov/fdsys/browse/collection.action?collectionCode=FR.*

[14] The *Manual of Regulation & Procedures for Federal Radio Frequency Management* can be found at *http:/www.ntia.doc.gov/osmhome/redbook/redbook.html.*

Exhibit 3-2: Relationship Between the FCC and NTIA

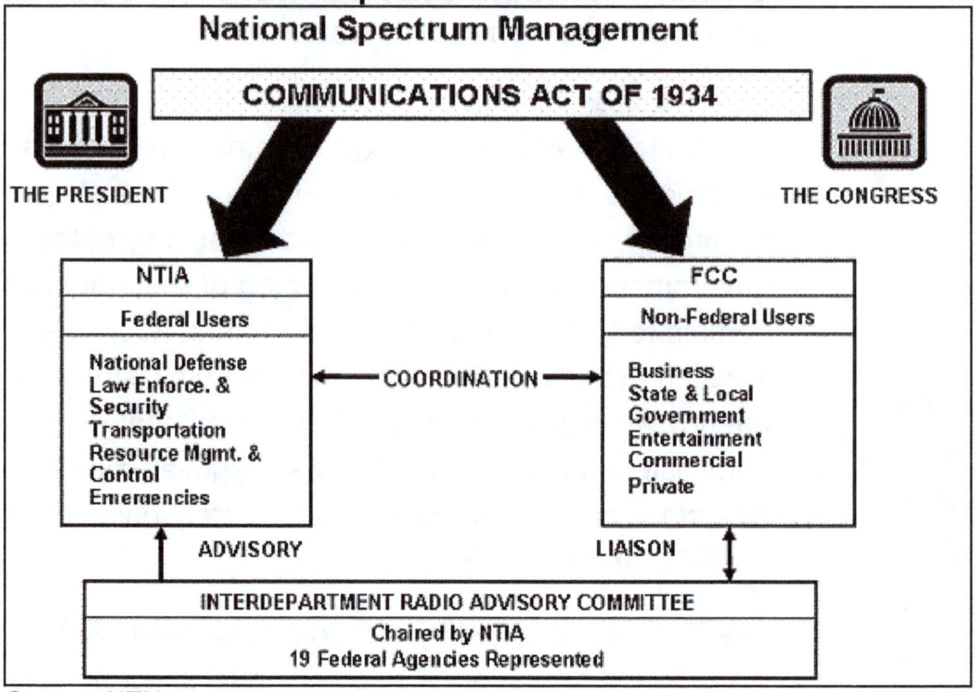

Source: NTIA.

The FCC mainly issues primary spectrum licenses to commercial entities, although a secondary market for these licenses has developed

In addition to the NTIA and the FCC, another government agency, the Nations Communication System (NSC) historically coordinated several emergency communications programs. The NSC was formed in the aftermath of the Cuban missile crisis in 1962 and includes 24 federal government departments and agencies. In July 6, 2012, President Obama signed an executive order to eliminate the NSC and send many of its functions to the Office of Emergency Communication within the Office of Cybersecurity and Communication.

2. Licensing Process

Before issuing licenses, the FCC allocates spectrum for a specific use. 47 C.F.R. 2.1 defines the permissible services for which the FCC can allocate spectrum. The FCC can later issue spectrum licenses for primary use and, to a lesser extent, for secondary use that allow license holders to use the frequency for services within the U.S. frequency allocation. Primary licensees have the right to use the spectrum within the terms of the license without interference from other users. Secondary licensees are allowed to use spectrum on limited terms that do not interfere with the primary licensee. Specifically, holders of secondary licenses:

> (i) Shall not cause harmful interference to stations of primary services to which frequencies are already assigned or to which frequencies may be assigned at a later date;
>
> (ii) Cannot claim protection from harmful interference from stations of a primary service to which frequencies are already assigned or may be assigned at a later date; and
>
> (iii) Can claim protection, however, from harmful interference from stations of the same or other secondary service(s) to which frequencies may be assigned at a later date.[15]

The licenses' terms dictate the arrangements of the spectrum assignments with respect to the geographic coverage area, the transmission characteristics and rules of service to prevent the interference of signals.

FCC licenses explicitly deny the licensees property rights in the terms of their licenses

Licenses for spectrum for two-way communication are often granted in two separate "paired blocks." One of the paired blocks is authorized for uplink and the other for downlink. Regulatory agencies do this because most two-way spectrum transmission technologies work better when the "uplink" (transmission from handset to the base station) is on a separate frequency from the "downlink" (transmission from the base station to a handset).[16] Typically, there is a space of spectrum between the uplink and downlink segments to prevent interference. This space is called the "duplex gap."

Significantly, the terms of FCC licenses explicitly deny the licensees property rights in the terms of their licenses. They are for limited periods of time, typically eight years for television broadcasters, ten years for mobile operators and fifteen years for

---

[15] 47 C.F.R. 2.105(c)(2).

[16] An exception is Time Division Duplexing (TDD) technology which divides the uplink and downlink by time as opposed to frequency. TDD is not used heavily in the US, but is dominant in China. See Section IV.A.5.B of this Handbook for an additional discussion of TDD technology.

satellite operators. However, the FCC wants to encourage license holders to invest heavily in both buying spectrum at auction and building their businesses. As a result, spectrum licenses have taken on many of the economic attributes of property including the right to sell.[17] License holders often invest heavily in acquiring licenses at auctions and building out services. This investment often requires an expectation of long-term control of the spectrum to justify the investment, making it politically difficult to deny renewal absent egregious abuses of the license terms.

The FCC issues licenses in specific geographical areas...These areas differ depending on the nature of the license and the auction

Despite the fact that the FCC prohibits security interests in licenses, investors in spectrum license holders have been able to circumvent the impact of ownership restrictions. They do this by laying claim to the "economic value" of the license, as opposed to the licenses themselves. Licensees typically affect an economic claim to FCC licenses by putting the license(s) in a holding company and using the stock of the holding company as collateral. In this case, security investors have a stake in the "economic value" and proceeds of licenses when sold. Bankruptcy courts have generally upheld this strategy when properly done.[18]

3. Licensing Areas Vary

The FCC issues licenses in specific geographical areas. FCC divides the country into geographical units for the auctions. These areas differ depending on the nature of the license and the auction. License areas can range from the whole country to one of 734 Cellular Market Areas (CMAs). This creates a confusing system of geographical units of license areas. These licensing areas are described in Appendix III.

---

[17] For a detailed discussion of property rights in FCC spectrum, see: J. Armand Musey, "Broadcasting Licenses: Ownership Rights and the Spectrum Rationalization Challenge," *Columbia Science & Technology Law Review*, Vol. 13, 2013 (September 10, 2012). Available at SSRN: http://ssrn.com/abstract=1952138.

[18] Ibid., 327-330.

4. Current Emphasis of FCC

In recent years, the FCC's focus has largely been to keep its regulations current with rapidly changing technology. New technology often undermines the logic behind previous rules. This unnecessarily limits choices available to consumers, service providers and equipment manufacturers. New technology can also create new spectrum needs. The corollary is also true – regulatory restrictions can drive technological development. However, the prevailing view of U.S. regulators is that, in general, society is best served when market demand drives technological development. Consequently the FCC seeks to update the regulatory framework to accommodate that demand. Currently, the FCC's biggest priority is allocating additional bandwidth to mobile broadband as part of the National Broadband Plan.[19]

The FCC, unlike the spectrum regulatory agencies in other countries, has focused its efforts almost entirely on transmission requirement as opposed to requirement of receivers of

The FCC, unlike other the spectrum regulatory agencies in other countries, has focused its efforts almost entirely on transmission requirements as opposed to requirements for receivers of transmissions. If receivers work beyond their allocated spectrum band into a neighboring band, a licensed operator in that neighboring frequency band often cannot use its spectrum without interfering with the non-compliant receivers. Once they reach a critical mass, it is often politically hard to move these non-compliant receivers. In the case of GPS, the widespread availability of receivers that received signals outside of their spectrum limited Lightsquared's ability to use adjacent spectrum.[20] Going forward, Summit Ridge Group expects the FCC to place additional emphasis on receiver equipment compliance.[21]

---

[19] For more information about the 2010 National Broadband Plan, see: http://www.broadband.gov/plan.

[20] For a summary of the Lightsquared/GPS interference issue, see: http://en.wikipedia.org/wiki/LightSquared.

[21] For a detailed discussion of the receiver regulation issue in the United States, see: U.S. General Accounting Office, "Spectrum Management: Further Considerations of Options to Improve Receiver Performance," February 2013. Available at: http://www.gao.gov/assets/660/652284.pdf.

5. Government Spectrum Use Questioned

The NTIA has shown limited ability to effectively manage federal spectrum use or to evaluate the actual federal needs for spectrum

Sixty percent of the most valuable U.S. spectrum has either exclusive or primary allocation to government users such as the Department of Defense.[22] According to a report released by the U.S. Government Accountability Office (GAO) in 2011, however, the NTIA has shown limited ability to effectively manage federal spectrum use or to evaluate the actual federal needs for spectrum. The NTIA has instead focused on preventing interference between spectrum allocated to different government agencies. Most observers believe the spectrum used by government entities is underutilized in many areas. However, due to lenient reporting requirements for the government regarding its spectrum use, the actual figures on available or underutilized spectrum remain opaque.[23] On June 14th, 2013, a presidential memo, among other things, directed the NTIA to assess the actual uses of spectrum assigned to federal agencies.[24]

F. Regulatory Challenges are Significant

1. Current Regulatory Process Contributes to "Spectrum Crunch"

Delays in government spectrum allocation create inefficiencies that are leading to a shortage of spectrum in the United States wireless industry commonly referred to as the wireless "spectrum crunch." Inefficiencies emerge when the market demand for spectrum changes, as recently evidenced by the increased demand for wireless broadband. FCC license policies, however, do not permit licensees to change their use to accommodate the market changes.

---

[22] "Realizing the Full Potential of Government-Held Spectrum to Spur Economic Growth," President's Council of Advisors on Science and Technology (PCAST), July 2012, 8. Hereinafter referred to in text as "PCAST Report."

[23] United States Government Accountability Office, "SPECTRUM MANAGEMENT: NTIA Planning Processes Need Strengthening to Promote the Efficient Use of Spectrum by Federal Agencies." GAO-11-352. April 2012, 15 GAO-11-352.

[24] The White House, *Presidential Memorandum – Expanding America's Leadership in Wireless Innovation, June 14th, 2013.* Available at: http://www.whitehouse.gov/the-press-office/2013/06/14/presidential-memorandum-expanding-americas-leadership-wireless-innovatio . Hereinafter referred to in text as "Presidential Memo."

After market changes indicate a higher value use for spectrum, licensees must often wait for years until the FCC allows them to use the spectrum for that new use. These regulatory delays are a frequent source of frustration among service providers. However, the U.S. FCC is generally considered more rapid in its policy changing process than its European and Asian counterparts.

Since 1994, the FCC has issued licenses almost entirely through auctions

Licensing for primary (priority) use account for the most significant portion of licenses issued by the FCC. Historically, spectrum licenses have been distributed either through applications or auctions. In recent years, governments around the world have increasingly turned to auctions to allocate spectrum licenses. This trend has been driven by the desire to avoid the perception of bias on the part of those judging applications and has improved the chances that licenses go to their highest use. Since 1994, the FCC has issued licenses almost entirely through auctions. This method of distributing spectrum rights has limited access to spectrum for many service providers because auctions are infrequent and time consuming. Accordingly, many service providers bid for spectrum in excess of their current needs to ensure a sufficient amount for future needs. Moreover, incumbent service providers are typically in a position to pay the most at auctions, thus limiting new competition. These characteristics result in heavy centralization and underutilization of the available spectrum, which most economists agree is worth many times what the government raises at auction in terms of economic benefit to society.[25]

Another problem with auctions is that entities needing smaller, more tailored licenses may not be able to justify bidding enough for the large license packages. In the United States, this problem is mitigated by the fact that most terrestrial licenses are allocated in

[25] See, e.g.: Thomas W. Hazlett & Roberto E. Munoz, A Welfare Analysis of Spectrum Allocation Policies, Geo. Mason L. & Econ. Res. Paper No. 06-28 at 2 (Jan.19, 2008), available at http://ssrn.com/abstract=908717 (arguing that total social welfare benefits produced in public spectrum auctions are dominant over auction revenues).

small regions as opposed to nationally as in many other countries.[26]

2. Solutions Are Politically Difficult

[T]his process [of reclaiming and reallocating spectrum] typically takes many years, depriving the pubic of the most efficient use in the interim

One solution to eliminating the spectrum inefficiencies has been for the FCC to reclaim the spectrum that was demonstrably underutilized or not utilized at all by the licensee and then to reallocate it for higher value uses. Unfortunately, this process typically takes many years, depriving the public of the most efficient use of the spectrum in the interim. Examples of the time required for the FCC to reallocate spectrum through administrative processes are shown in Exhibit 3-3 below. Another manner to eliminate these efficiencies would be to allow existing license holders to use the spectrum for the new higher use. But, governments often find this solution to be impracticable. For instance, an existing licensee who may have received a license at little or no cost, would be allowed to have the rights to use it for a more valuable new service without paying for the "upgrade." This distorts the market because the upgraded license holder is free to compete against companies who may have paid billions of dollars for their licenses. The mobile wireless companies, for example, complained bitterly when the satellite telephony companies were given permission to use their spectrum terrestrially at no additional cost.[27]

---

[26] For a detailed critique of the spectrum auction process, see: Thomas W. Hazlett, Roberto E. Muñoz, and Diego B. Avanzini, "What Really Matters in Spectrum Allocation Design," 10 NW. J. TECH. & INTELL. Prop. 93 (2012), http://scholarlycommons.law.northwestern.edu/njtip/vol10/iss3/2.

[27] For a summary of the wireless industry's complaints about satellite ATC, see: Steve Lovelady, "Limits Laid for Land-Based Sat Service Systems, available at http://www.fhhlaw.com/resources/TelecomLaw/2005/april/0405tl_8.pdf.

Exhibit 3-3: Time Historically Required to Reallocate Spectrum

Band	First Step	Available for Use	~ Lag Time
Cellular (AMPS)	1970	1981	11 years
PCS	1989	1995	6 years
EBS/BRS	1996	2006	10 years
700 MHz	1996	2009	13 years
AWS-1	2000	2006	6 years

Source: FCC and Summit Ridge Group, LLC Analysis.

3. Secondary Spectrum Market Improves Efficiency

Over the past 10-15 years, the FCC has gradually eased barriers to buying and selling spectrum rights

Over the past 10 to 15 years, the FCC has gradually eased barriers for buying and selling spectrum rights. In 2000 the FCC revised its spectrum policy in the report "*Principles for Promoting the Efficient Use of Spectrum by Encouraging the Development of Secondary Markets.*"[28] This new market-based policy, which was further defined and extended in subsequent reports in 2002 and 2004,[29] suggested establishing a secondary market for licenses to create a more efficient and dynamic market for spectrum licenses.

This market serves as a platform for breaking the larger frequency licenses acquired on the primary market into smaller, more differentiated licenses that are then subleased to smaller entities. This reduces inefficiencies by creating cheaper licenses that can be more customized to the needs of a secondary buyer. These subleases determine the rights under which the primary spectrum holder (the lessor) can provide another entity (the lessee) access to certain parts of its spectrum in accordance with the FCC determined standards. Despite the regulatory progress, the lack of available hardware capable of dynamically changing frequency presents a major challenge to spectrum leasing. Nonetheless, recently developed smart radio technology, operable on a wide range of bands, may ameliorate this problem.

---

[28] FCC Policy Statement (FCC 00-401). December 1, 2000, available at: http://hraunfoss.fcc.gov/edocs_public/attachmatch/FCC-00-401A1.pdf.

[29] For a list of FCC documents related to its secondary market policies, see: The Secondary Markets Initiative on the FCC's website. Available at: http://wireless.fcc.gov/licensing/index.htm?job=secondary_markets.

4. Spectrum Transfer Approval Process More Streamlined

Despite the liberalizat-ion of the rules for buying and selling spectrum, the FCC has not conceded the ultimate government ownership of spectrum

Spectrum transfers in the secondary market generally occur in one of two forms: a de facto transfer or a spectrum manager lease. De facto transfers are FCC approved arrangements in which the spectrum holder (licensee) effectively transfers the control of a part of the spectrum to the lessee for a specified period (short-term leases for under 360 days, long-term leases for over 360 days). Meanwhile, de jure (rightful) control of the spectrum remains with the licensee. The second form of transfer, the spectrum manager lease, is an arrangement that allows the licensee to provide the lessee with a certain range of frequencies as long as de facto and de jure control over that particular spectrum remain with the original spectrum holder. Since de facto control is not transferred, no approval from the FCC is needed in this process.

The processes of acquiring spectrum licenses were defined and streamlined by the reports the FCC published in 2000, 2002 and 2004.[30] However, there currently is no existing market-making intermediary that facilitates the trading of the licenses between spectrum holders (licensees) and spectrum users (lessees), and the FCC has not determined what entity would act as such. But, the FCC did state that the agency itself would not act in this role. The role would be most effectively filled by independent parties that compete with one another.[31]

Despite the liberalization of the rules for buying and selling spectrum rights and the streamlined approval process, the FCC has not conceded ultimate government ownership of spectrum. As such, until the government gives up this ultimate power, any

---

[30] The net effect of these reports is that the acquisition process has been streamlined. The general review process has been reduced from 30 to 21 days, while certain de facto transfer proposals could be reviewed as fast as within a single business day. See FCC 04-167, Second Report and Order, Order on Reconsideration, and Second Further Notice of Proposed Rulemaking, September 2004, 19.

[31] Independent parties that have risen to service this need include Spectrum Bridge, Inc. (see: http://spectrumbridge.com) and Telcordia Technologies, a division of Ericsson (see: http://www.ericsson.com).

change of control will require its approval .

5. Economists Support Arguments for Privatization

In 1959, the economist Ronald Coase argued that privatizing spectrum ownership would result in a rapid reallocation to the most productive use for society

Advocates of the market-based approach of allocating spectrum have campaigned for the privatization and free sale of spectrum, without the need for any license. One of the most notable advocates was Nobel Prize winning economist Ronald Coase. In 1959, he argued that privatizing spectrum ownership would result in a rapid reallocation to the most productive use for society.[32] But, governments around the world have resisted selling spectrum. They see it, to some extent as a public good rather than a exchangeable commodity. Additionally, some academic research points to problems that can arise from unrestricted market trading of spectrum. These include spectrum hoarding and excessive fragmentation of spectrum, which impedes the formation of useful units.[33] Nevertheless, the licensing regimes of many countries, including the United States, go a long way towards mimicking property rights without explicitly granting them.

6. Prime Spectrum Is Used Inefficiently

Despite the current crunch, spectrum is underutilized

Despite the current crunch, spectrum is underutilized. Even in congested urban areas, prime spectrum below 3.7 GHz is used at less than 20% of capacity.[34] This is due to the lack of incentives for efficient use by the FCC and the lack of market-based systems to

---

[32] Ronald Coase, "The Federal Communications Commission," 2 J.L 2 J.L. & Econ. 1 (Oct. 1959).

[33] For a detailed academic discussion of the potential challenges related to unregulated spectrum markets, see Patrick Xavier & Dimitri Ypsilanti "Policy Issues in Spectrum Trading", Vol. 8 Issue 2, 2006, 34-61. Available at: http://www.emeraldinsight.com/journals.htm?articleid=1546218&show=abstract.

[34] PCAST Report at 99 citing: R.B. Bacchus K.J. Zdunek, & D.A. Roberson, "Long-term Spectrum Occupancy Findings in Chicago," 2011, IEEE Symposium: New Frontiers in Dynamic Spectrum Access Networks. dx.doi.org/10.1109/DYSPAN.2011.5936195; Mark McHenry, "NSF Spectrum Occupancy Measurements: Project Summary," Shared Spectrum Company, 2003, 2005, available at www.sharedspectrum.com/measurements; Mark McHenry & M. and M. Vilimpoc, "Dupont Circle Spectrum Utilization During Peak Hours: a Collaborative Effort of the New America Foundation and the Shared Spectrum Company," A New America Foundation Issue Brief.

encourage spectrum sharing among commercial users.[35]

7. Unlicensed Spectrum Gaining Momentum

In the past few years, there has been an increasingly large movement towards the use of unlicensed spectrum. Unlicensed spectrum has been in use for decades for devices like television remote controls and garage door openers. However, the massive success of Wi-Fi and the development of "smart radios" that can change frequencies as needed to avoid interference have increasingly caused industry observers to see unlicensed spectrum as an alternative to licensed spectrum. As detailed in Section V.B below, the FCC and the NTIA are currently evaluating a significant expansion of unlicensed spectrum use, including the current processes at 3.5 GHz and 5 GHz. Unlicensed spectrum is likely to play a significant role in the dialogue over the resolution of the "Spectrum Crunch."

... the influx of additional usable spectrum may also put downward pressure on the value of spectrum

Various proposals, including the PCAST Report, have encouraged governments to implement systems to share unused spectrum between government and commercial users. Some European countries have begun similar initiatives. These initiatives, if successful, may significantly expand available spectrum and allow for growth of additional wireless-based services. However, the influx of additional usable spectrum may also put downward pressure on the value of spectrum. Section VI.H below discusses this point in greater depth.

8. Effect of Lobbying Efforts

The wireless industry invests roughly $30 billion annually in capital expenditures to build out networks. The major carriers average $5 billion to $10 billion in capital expenditures each. Given that the industry is heavily regulated, it would be irresponsible for them to

---

[35] International approaches to developing efficiencies in government spectrum vary. In the United Kingdom, for example, the military was required to value the spectrum that it uses. If private entities are willing and able to meet this price, the military is required to move from the spectrum, presumably using the funds from the private sector to do so.

not protect their investment with appropriate political lobbying efforts. It has been reported that the Big-4 and the CTIA spent over $53 million in lobbying efforts in 2012 (www.opensecrets.org) and occupy three spots of the list of the 30 largest spenders on lobbying. This is disproportionate to the industry's size in the economy, but perhaps appropriate due to the regulatory intensiveness of the industry. Their lobbying budget allows the wireless industry to maintain constant vigilance on pending legislation and FCC policy changes.

The wireless industry uses its lobbying budget to maintain constant vigilance on pending legislation and FCC policy changes...

The wireless industry uses its lobbying budget to maintain constant vigilance on pending legislation and FCC policy changes and to submit comments on almost every docket (e.g., AT&T filed 11 comments on the broadcast incentive auction process alone). This gives the wireless industry significant access to members of Congress, affording it influence over spectrum policy. For example, when the FCC attempted to prevent AT&T and Verizon's participation in the broadcast incentive auctions, Congress quickly responded with legislation instructing the FCC not to prevent any qualified bidder from participating.[36]

The power of a few strong industry players often sets the narrative of how an issue is explained. The major wireless industry players have stressed the "spectrum crunch," with little supporting information, as the basis for their need for significant amounts of additional spectrum. Only recently have opponents challenged the data and assumptions behind the spectrum crunch hypothesis due to their disadvantage in funding. But they also lack conclusive information and are years behind in the public/political debate.

[36] The FCC is able to set spectrum caps and other limits on auction participants. AT&T and Verizon are lobbying to prevent this from being part of the FCC's final auction rules.

PART TWO: ECONOMIC AND REGULATORY DYNAMICS

IV. Major Commercial and Policy Spectrum Issues

A. General Spectrum Dynamics

1. The Looming Spectrum "Crunch"

The spectrum crunch refers to a scarcity of spectrum available to commercial entities

Spectrum is a finite resource. The mobile industry's need for spectrum is rapidly growing (see Exhibit 4-1 below for projected wireless traffic growth). But government and major commercial entities control the majority of spectrum—many of whom use their acquired frequencies inefficiently. Only recently have the significant economic and social benefits of quality broadband service been recognized.[37] Governments around the world are now moving to allocate more spectrum for this purpose. The slow nature of government policy change, however, is clashing with the rapid demand for wireless data. The potential inability to meet prospective spectrum demands is known as the "spectrum crunch."

Exhibit 4-1: Projected Mobile Data Traffic Growth

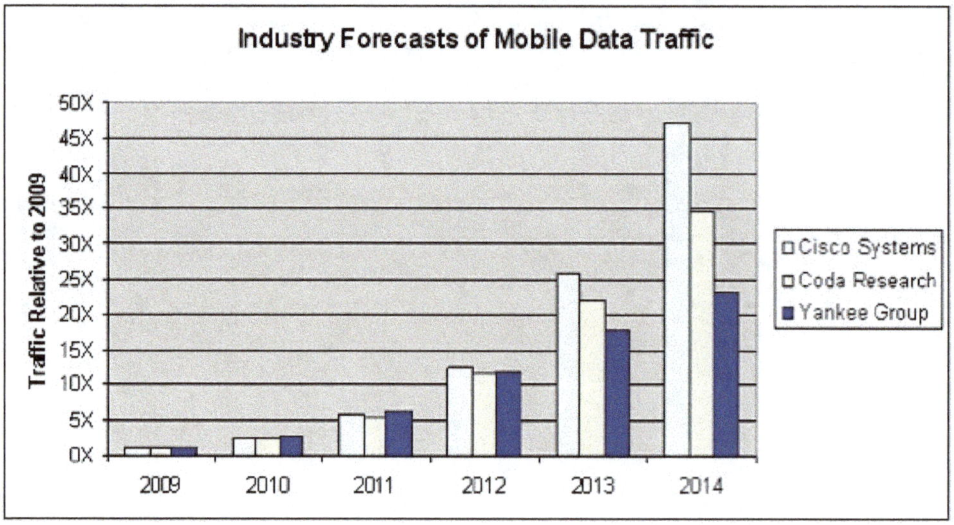

Source: FCC; National Broadband Plan.

[37] Hazlett, Thomas W. and Munoz, Roberto E., A Welfare Analysis of Spectrum Allocation Policies (January 19, 2008). George Mason Law & Economics Research Paper No. 06-28. Available at SSRN: http://ssrn.com/abstract=908717 or http://dx.doi.org/10.2139/ssrn.908717

While the total supply is finite, new technology has increased spectral efficiency enabling service providers to offer more data with existing spectrum. But, achieving incremental spectral efficiency improvements is limited. Accordingly, wireless services providers need access to additional spectrum to meet increasing demand requirements. The federal government owns approximately 60 percent of the prime U.S. spectrum supply. Most observers consider its use of this spectrum to be highly inefficient. For the needs of mobile wireless demand to be met, the government must reallocate or share some of its inefficiently used frequency bands with commercial entities.[38]

There is much uncertainty over both demand growth for wireless services and the amount and timing of inefficiently used spectrum becoming available for mobile broadband services. This dual uncertainty has resulted in vast differences in the perceived value of spectrum. Nonetheless, major wireless providers around the world are clamoring for additional spectrum to meet their customers' needs. Of course, any spectrum allocated to the existing industry limits the spectrum available for new entrants to share spectrum.

Spectrum licenses are granted on a fundamental element of property theory... If you cannot exclude others, then the "free rider" effect will prevent a return on investment to the builder

2. Spectrum Sharing Movement and "White Spaces"

A movement is growing to encourage governments to leave large amounts of spectrum for unlicensed use by anyone with approved hardware.

a. End of Property Theory?

Spectrum licenses are granted on a fundamental element of property theory: one needs exclusive rights to property (i.e. licenses to spectrum) to economically justify development. If you cannot exclude others, then the "free rider" effect will prevent a return on

---

[38] Commissioner Ajit Pai (Diary), "Too Much Government, Too Little Spectrum," January 3, 2013, accessed May 15, 2013, http://www.redstate.com/ajitpai/2013/01/03/too-much-government-too-little-spectrum.

investment to the builder. This is a fundamental justification for private property law – investors must be able to exclude others or no one will invest in projects on that property. However, a small but growing movement is advocating government allocation of large amounts of unlicensed spectrum for shared use.

Advocates of this movement suggest that, at a minimum, unused licensed broadcasting spectrum, called "white spaces," should be open for public use if the licensee is not using it. Shared use should be possible without the need for licenses. The spectrum sharing advocates argue that modern technology, particularly software defined "smart radios," allow for increased spectrum efficiency by enabling unused spectrum to be shared. Smart radios allow users to dynamically "hunt" for and identify open frequencies as needed while sharing the available "white spaces" of spectrum. The argument is that if the government allows current spectrum to be fully utilized via spectrum sharing, spectral efficiency would increase and the spectrum crunch would greatly diminish or even disappear. Smart radio technologies can access remote databases[39] to avoid using emergency or other critical spectrum at different frequencies in different areas. Some shared spectrum advocates go as far as to say that the property right model has reached the limit of maximizing spectrum efficiency and that new sharing models are needed to achieve greater efficiencies.

[T]he FCC has allowed some "white spaces" found in commercial licensed spectrum bands to be used by unlicensed users under certain terms.

In response to the "white spaces" movement, the FCC has allowed some "white spaces" found in commercial licensed spectrum bands to be used by unlicensed users under certain terms. These terms include shorter time periods and lower costs for the commercial users subleasing the spectrum. They are intended to encourage smaller firms to join the spectrum market and spur economic growth, creating jobs. One specific spectrum sharing strategy suggested by the PCAST Report urges for the clearing of broad

---

[39] Spectrum Bridge and Telcorida Technologies are two companies which operate such databases to allow software-defined "smart" radios to operate within approved spectrum.

frequency bands from 2700 to 3700 MHz for shared use by users that operate better on broader rather than narrower spectrum bands. This would facilitate shared low-power, small cell operations and increase efficiency within the commercial spectrum. The FCC is currently evaluating several bands for shared use including ongoing NPRM processes at the 1755 MHz, the 3.5 GHz, 4.9 GHz and 5.9 GH bands as shown in Exhibit 5-9 below.

One concern about the sharing movement is that it might prevent wireless operators from meeting service commitments to customers. Some users could dominate shared spectrum at a given time and place, preventing others from obtaining needed capacity. In this situation, the individual operator would not have the ability to manage traffic on the network as it would if it owned the spectrum itself. This concern has also hampered fundraising efforts by unlicensed service providers. For these reasons, established wireless operators have indicated a clear preference for licensed spectrum.

b. Starting with Government Spectrum?

In addition to commercial spectrum, the spectrum sharing movement also targets portions of under-utilized U.S. government held spectrum. One challenge is that some government spectrum is for emergency purposes. Although seldom used, the government requires access to it. Many government agencies are fearful that if they share their spectrum with unlicensed users, they may not have adequate access to it during an emergency. However, there is a precedent for such sharing. Verizon is required to share its D Block in the upper 700 MHz band with the adjacent future public safety network. Effectively Verizon is a secondary user of this spectrum and the public safety network has priority in case of an emergency (see Section IX.A.1 below for a discussion of this band plan).

Another issue with applying spectrum sharing to government held spectrum is the lack of incentive for the government to improve its spectral efficiency

Another issue with apply spectrum sharing to government held spectrum is the lack of incentives for the government to improve its spectral efficiency. While the commercial sector must report and justify its use of spectrum, as well as satisfy investors' expectations of optimal asset utilization, the federal government has little accountability to keep its spectrum. This makes it unclear which pieces of the federal spectrum are used inefficiently and to what extent. As long as there is no common standard for determining commercial and government spectrum efficiency, it is unlikely the government will optimize its use of spectrum. The PCAST Report suggested that a "currency-like" accounting allocation[40] and incentive system be put into place to create a long-term motivation to manage and share government spectrum better. The recent Presidential Memo on spectrum sharing may put additional pressure on federal agencies to share spectrum.

3. Narrowbanding

In addition to the digital television transition, the FCC has sought to free additional spectrum by taking advantage of new technology that allows for greater spectral efficiency. In certain bands, it requires users to switch to equipment that uses less spectrum. As of January 1, 2013, public safety and industrial/business land mobile radio systems using the 150 to 174 MHz and 421 to 470 MHz bands were required to switch from 25 kHz technology to 12.5 kHz or better technology. A similar narrow-banding plan is under evaluation for the 470 to 512 MHz band (the T-Band).

New wireless technologies [that] have the ability to get more information through a given amount of bandwidth ...have played a significant role in delaying the spectrum crunch

4. Advanced Technologies are More Efficient

New wireless technologies have the ability to get more information through a fixed amount of bandwidth. Such advances have played a significant role in delaying the spectrum crunch, but it is unclear if they can continue indefinitely. They should slow as we reach the physical limits of spectrum utilization. Exhibit 4-2 below graphically illustrates the improvements in spectrum utilization with major wireless technologies.

---

[40] PCAST Report, 55.

Exhibit 4-2: Downlink Bps/Hz by Spectrum Technologies[41]

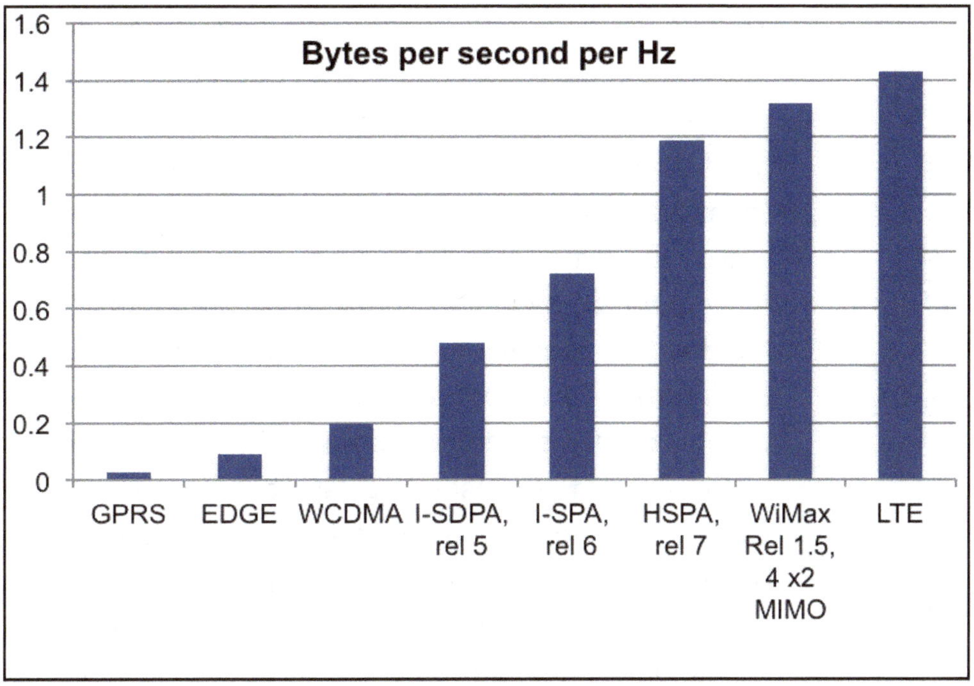

Source: FCC and Summit Ridge Group, LLC analysis.

Smaller than ideal blocks of spectrum and more users in a cell site degrade the performance as do interference factors such as physical obstructions (buildings)

The ability for the technologies to achieve the throughput performance indicated above is predicated on a number of factors. Smaller than ideal blocks of spectrum and more users in a cell site degrade performance as do interference factors such as physical obstructions like buildings.

5. Broadband and LTE Technologies May Change Spectrum Needs

Broadband and LTE Technologies may significantly change the nature of spectrum needs.

---

[41] 1 byte = 8 bits. The chart only displays the GSM/3GPP family of technologies. Performance of EV-DO standards is comparable with HSPA. See Letter from Dean R. Brenner, Vice Pres., Gov't Aff., Qualcomm Inc., to Marlene H. Dortch, Secretary, FCC, GN Docket No. 09-51 (Dec. 9, 2009). Figure shows downlink capacities calculated for 2x10MHz spectrum availability. Estimates of spectral efficiency calculated for each technology with the following antenna configuration: WCDMA, 1x1 and 1x2; HSPDA, Rel.5, 1x1; HSPA Rel. 6, 1x2; HSPA, Rel. 7, 1x1 and 1x2; LTE, 1x1 and 1x2. Information in this footnote is from the FCC – National Broadband Plan.

a. More Downlink Spectrum Needed

The FCC is considering fixing asymmetry as well as creating flexible band plans that allow for downlink expansion

Traditionally wireless services had symmetrical demand for spectrum in each direction of use, because each person in a voice conversation communicates about the same amount. As a result, most band plans allocated spectrum in equal sized pairs for uplink (from the handset to the base station) and downlink (from the base station to the handset). However, as data traffic and media viewing has grown, demand for capacity from the Internet down to the handset has greatly outpaced upward demand from the handset to the Internet. Alcatel's Stephen A. Wilkus recently estimated that downlink traffic exceeded uplink traffic by a 4:1 ratio in 2005. By 2011, this had increased to a ratio between 17 to 30:1.[42,43] Downlink traffic is about twice as spectrally efficient as uplink traffic, offsetting some, but not all of this difference. As a result of the higher demand for downlink spectrum, the future need for spectrum allocation in equal sized pairs is in doubt. However services providers have been reluctant to advocate changes that would require them to redesign their networks. To address asymmetrical usage, the FCC is considering non-paired spectrum allocations as well as creating flexible band plans that allow for downlink expansion on paired spectrum. An illustration of asymmetrical bandplans is shown in Exhibit 4-3 below.

---

[42] Stephan A. Wilkus, "TDD and Asymmetrical FDD" presented at the FCC Band Plan Technical Forum, July 12, 2012, available at: http://transition.fcc.gov/bureaus/oet/tac/tacdocs/meeting71612/PANEL2.2-Wilkus-Alcatel-Lucent.pdf.

[43] Al Jette, estimates the ratio of downlink to uplink traffic at between 6 and 13 to 1. See Al Jette, "FCC Forum on the Future of Wireless Bands, at 2-3, FCC Forum on the Future of Wireless Bands, July 16, 2012, available at: http://transition.fcc.gov/bureaus/oet/tac/tacdocs/meeting71612/PANEL2.1-Jette-NokiaSiemensNetworks.pdf

Exhibit 4-3: Asymmetrical Band Plans

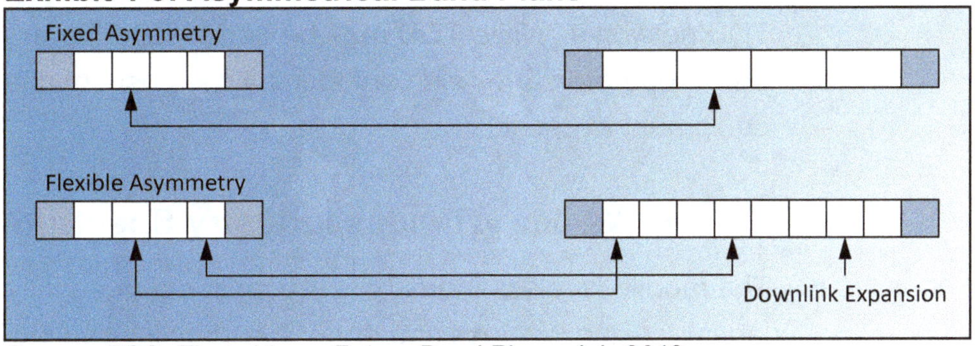

Source: FCC: Thoughts on Future Band Plans, July 2012.

b. FDD vs. TDD LTE Band Planning

Long Term Evolution (LTE) comes in two versions: Frequency Division Duplexing (FDD) and Time Division Duplexing (TDD). FDD, the more common version in the United States, uses two paired sets of frequency blocks one for the uplink and one for the downlink. TDD uses one frequency block for both uplink and downlink but segments them by time to avoid interference. This often makes TDD better suited for asymmetrical traffic. Clearwire is the only US operator using the LTE-TDD standard. Large international operators such as China Mobile use TDD Handsets with chipsets capable of handling both FDD and TDD. These handsets are expected to become widely in the coming years. An illustration of FDD and TTD band is shown in Exhibit 4-4 below.

Exhibit 4-4: FDD and TDD Frequency Usage

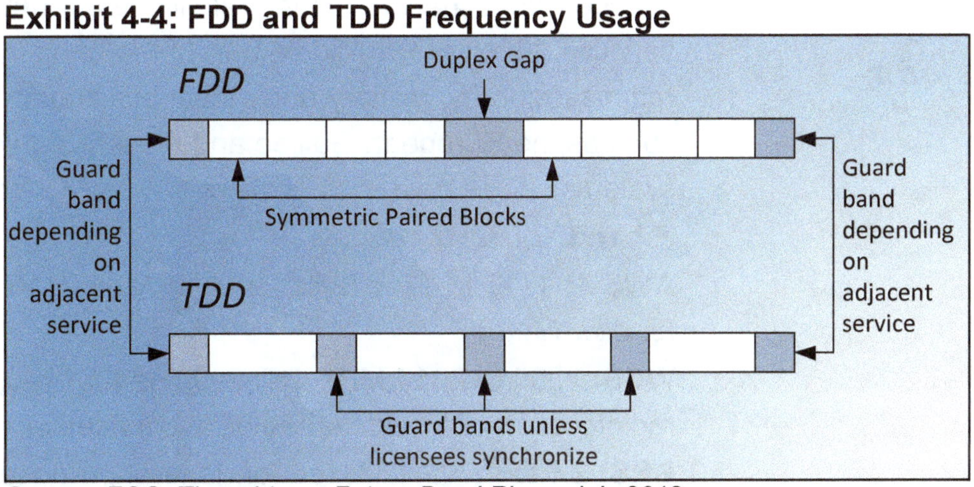

Source: FCC: Thoughts on Future Band Plans, July 2012.

In many bandwidth planning situations, regulators will need to decide between creating band plans optimized for either FDD or TDD networks. While TDD may be better suited to asymmetrical traffic, operators are reluctant to incur the costs of changing equipment to use it.

B. Mobile Wireless Industry Spectrum Dynamics

The mobile wireless industry is the most commercially important industry using wireless spectrum. It is growing rapidly in almost every country. Consequently, finding sufficient spectrum for mobile broadband uses large portions of the resources of spectrum regulators worldwide.

1. Bands Used

The mobile industry uses a patchwork of frequencies in the following bands between 700 MHz and 2.7 GHz:

Mobile wireless services use a patchwork of frequencies between 700MHz and 2.7 GHz.

- **700 MHz**
- **AMPS:** 800 MHz
- **PCS:** 1850 to 1910 MHz for uplink and 1930 to 1990 MHz for downlink.
- **L-band:** 1.6 GHz band. Satellite broadcasters are allocated 1482.352 to 1490.624 MHz and 1525 to 1646.5 MHz. The former is used primarily for satellite radio and related services while the latter is for satellite telephony (MSS) services. Much of the middle part of the band is reserved for radio astronomy, military and global positioning systems such as the European Galileo and the Russian GLONASS systems.
- **S-band:** 2.0 GHz band.
- **AWS-1:** 1710 to 1755 MHz for uplink and 2110 to 2155 MHz for downlink.
- **WCS:** 2305 to 2310 MHz. Primary use for fixed, mobile, radiolocation or broadcast-satellite services.
- **BRS:** 2.5 to 2.7 GHz. Used for general-purpose networking. This is where Clearwire holds most of its spectrum and where Sprint has it 4G network.

(Detailed analysis of mobile wireless spectrum bands can be found in Section IX of this Handbook)

2. Managing Growth in Demand is Difficult

Mobile wireless services are new, compared to services such as AM radio and television, but have grown rapidly. In the late 1980s analog mobile wireless services were offered in the 800 MHz range. As demand grew and the wireless services began to offer digital services, the FCC allocated additional spectrum to mobile wireless services, starting with the PCS spectrum between 1850 and 1990 MHz. Since 2006 the AWS (2 GHz band) and EBS/BRS (2.5 GHz) spectrum have also been allocated to mobile wireless services. Since 2006, spectrum available for these services has grown from 170 MHz to 608 MHz (see Exhibit 4-5 below).

Cisco estimates mobile data traffic will grow at a compound annual growth rate of 66% between 2012 and 2017.[44] Managing this growth has been a challenge for the FCC. Some industry observers expect growth in additional services such as Machine-to-Machine communications.[45]

---

[44] Cisco, "Cisco Visual Networking Index: Global Mobile Data Traffic Forecast Update, 2013-2017," 5 (Hereinafter referred to in text as *Cisco Visual Networking Index*).

[45] Jose Del Rosario, The Bottom Line, "The Cellular Threat to SCADA/M2M," January 8, 2013. Accessed on May 23, 2013, http://www.nsr.com/news-resources/the-bottom-line/the-cellular-threat-to-satellite-scadam2m/

Exhibit 4-5: Significant Growth in Mobile Spectrum

Band	Before 2006	Added Since 2006	Total
Cellular	50 MHz		50 MHz
PCS	120 MHz		120 MHz
SMR		14 MHz	14 MHz
700 MHz		70 MHz	70 MHz
Add'l PCS		10 MHz	10 MHz
AWS-1		90 MHz	90 MHz
AWS-4		40 MHz	40 MHz
WCS		20 MHz	30 MHz
EBS/BRS		194 MHz	194 MHz
Total	**170 MHz**	**438 MHz**	**608 MHz**
% Increase since 2006:			358%

Source: FCC and Summit Ridge Group, LLC analysis.

A recent analysis by Oppenheimer & Co. estimates that approximately 250 MHz of the 414 MHz of mobile spectrum below 2.5 GHz has been built out.

A recent analysis by Oppenheimer & Co. estimates that approximately 250 MHz of the 414 MHz of mobile spectrum below 2.5 GHz has been built out.[46] To further increase mobile spectrum, the FCC has begun to move towards allowing Mobile Satellite Service (MSS) spectrum holders to use their spectrum terresterially without handsets that also operate through the satellite. This essentially allows MSS spectrum license holders to use their spectrum like traditional mobile spectrum.

There are two primary strategies to overcome spectrum scarcity. One is to increase the underlying spectrum capacity. The other strategy is to improve the use of the available spectrum.

Innovations in technology have been a significant driver of efficiency in spectrum use. Technological developments, such as multi-antenna signal processing (MAS) and the digitization of wireless services have allowed certain services to be provided in the same way while using less spectrum. Spectrum requirements can be reduced due to technical improvements that reduce interference and improve compression, among other factors. When

[46] Tim Horan, Oppenheimer & Co. "Spectrum White Paper," March 7, 2013.

the FCC changed its policy in 2008 to allow wireless services providers to switch their analog services to digital services, previously used spectrum became available for new purposes since the transmission of digital signals requires less spectrum than that of analog signals many countries still use. The transition of television broadcasting to digital (discussed in depth in Section IX.C) has also facilitated repacking television stations and provided significant additional spectrum for mobile wireless service providers. This is often referred to as the "Digital Dividend."

If an entity only makes use of its leased frequency range at certain times, the spectrum should be able to be subleased to another party at other times to increase efficiency

3. Cell-Splitting Has Limits

Splitting mobile wireless cells into smaller areas greatly increases frequency reuse, which, in turn, increases capacity for wireless operators. However, cell splitting has limits. As cells get smaller, placement options for antennas become more limited. As cell sites are reduced to below a few hundred meters, the number of possible locations for antennas for a particular cell declines. Moreover, these locations may not be available in certain urban environments leaving "holes" in coverage that frustrate consumers. Currently the limits of cell splitting is not a major issue in the U.S., which has cell site density approximately 1/5 of the level found in Japan.

4. Spectrum Sharing May Increase Usage Efficiency

The spectrum sharing movement advocates the second method of coping with the mobile wireless spectrum crunch by opening up under-utilized spectrum to entities that do not need to acquire licenses to use this spectrum. The idea is that if an entity only makes use of its leased frequency range at certain times, the spectrum should be able to be subleased to another party at other times to increase efficiency. Spectrum sharing is more difficult to implement since it requires policy changes to determine the technical conditions under which unlicensed users can access spectrum without interfering with the licensed users who have priority. This will impact both commercial entities and the government.

The U.S. government itself supports the idea of sharing spectrum. As discussed by the 2012 PCAST report, the government believes that sharing spectrum will make spectrum abundant by multiplying the capacity of this commercial spectrum by a factor of 1,000.[47]

5. Offloading is a Significant Strategy

Wi-Fi is the most common form of spectrum sharing. Wireless operators are increasingly using Wi-Fi and femtocell offloading to fixed networks as a major spectrum management strategy. Cisco estimates that in 2012, 33% of global wireless traffic was offloaded via Wi-Fi and femtocells, and that offloading will increase to 71% in 2017.[48]

As the demand for spectrum increases in urban areas, wireless companies will seek to add higher frequency spectrum in urban areas where it is more readily available

Offloading is sometimes controversial with consumer groups. While it saves carriers money, it is a drain on cell phones' battery life and is sometimes done without the knowledge of the consumer. When carriers offload to open networks, it can compromise consumers' security. Summit Ridge Group expects additional regulation concerning Wi-Fi offloading practices in the coming years. We also expect the trend toward greater Wi-Fi offloading to increase.

6. Spectrum Portfolio Diversity Becoming Increasingly Important

As the demand for spectrum increases in urban areas, wireless companies will seek to add higher frequency spectrum in urban areas where it is more readily available. In some respects this spectrum is better suited for small cell sites in urban areas, but it suffers from weaker building penetration. An analysis by Nokia suggests that indoor penetration at 900 MHz can be 60% to 130% greater than at 2100 MHz.[49] Despite this limitation, higher frequencies are critical to providing sufficient capacity in densely populated high use areas. Wireless operators also need to cover rural areas where large cell sites provide the most economical way to cover vast areas. In rural areas, spectrum below 1 GHz is most desirable. To accomplish these goals wireless operators need a diverse portfolio of spectrum to handle the environments that they

---

[47] PCAST Report, vi.

[48] *Cisco Visual Networking Index*, 11-12.

[49] Nokia Siemens Networks, "WCDMA Frequency Refarming," available at: http://www.nokiasiemensnetworks.com/portfolio/solutions/wcdma-refarming.

serve.

7. Interoperability Becoming a Significant Issue

Given the proliferation of spectrum bands and the increasing cost of making advanced handsets..., [hardware] interoperability...is becoming a significant issue

Given the proliferation of spectrum bands and the increasing cost of making advanced handsets capable working on multiple bands, interoperability (handsets work on all service providers' spectrum) is becoming a significant issue. In 2012, the FCC issued a NPRM for the lower 700 MHz band indicating its desire that all handsets made for the lower 700 MHz band work on all channels within the lower 700 MHz band. This was particularly important to licensees of the 6 MHz in the band's A Block of spectrum. Licensees of this block cannot use the full 6 MHz due to a requirement not to interfere with the adjacent television channel 51. As a result, smaller wireless operators typically bid for the more affordable A Block at auction. Licensees of the more attractive spectrum blocks had little incentive to ask equipment manufacturers to make handsets to work also on this block. And equipment manufacturers did not have the economies of scale to make handsets at attractive prices for smaller operators on this block. As a result valuable spectrum was underutilized. Hence, the FCC stepped in with the RM-11592 to require interoperability so that handsets would be able to work on the full 700 MHz band, thus maximizing its use. As more spectrum becomes available, similar interoperability issues are likely to arise.

8. Thoughts on Valuation

Generally speaking, the lower frequency mobile wireless bands are the most valuable, due to their greater coverage. This is particularly true if the band is unencumbered and has equipment available for it. Summit Ridge Group expects the future value of higher frequency spectrum, such as the 2.5 GHz band to depend on the rate at which the FCC allocates additional spectrum and/or sharing technologies to increase capacity. Industry-wide, there is little consensus on the rate of future growth of either mobile data demand or spectrum supply.

C. Unlicensed Spectrum/Spectrum Sharing

The challenge of allocating sufficient licensed spectrum to mobile broadband has lead to new ideas. The most prominent of these is shared spectrum. The FCC has allocated several bands for low-powered unlicensed services. These are particularly common for devices using small amounts of bandwidth on a short-term basis over short distances. Such devices include heart monitors, garage door openers, remote-control devices for household appliances. Generally, these uses are uncontroversial since the limited transmission range and power avoids interference, and spectrum capacity limits are rarely an issue. However, governments around the world are examining the prospect of making additional unlicensed (shared) spectrum available for mobile broadband through Wi-Fi and other technologies. These new broadband applications raise more complex interference issues. While the debate is far from settled, the emerging consensus in the industry is that interference issues are manageable with modern technology. Consequently, the debate is now turning to how to manage interference in shared spectrum and how conservative the rules should be to protect the rights of priority users.

...the emerging consensus in the industry is emerging that [spectrum sharing related] interference issues are manageable with modern technology

1. Bands Allocated for Unlicensed Broadband

The following bands are used for spectrum sharing:

- TV White Spaces
- 902-928 MHz (Wi-Fi – 26 MHz)
- 1880-1930 MHz
- 2400-2483.6 MHz (Wi-Fi – 83.6 MHz)
- 3550-3700 MHz
- 5150-5350 & 5470-5825 MHz (Wi-Fi – 555 MHz)

Wi-Fi, which uses shared spectrum, has been, an enormous success. This shared spectrum success of Wi-Fi as well as the U.S. government's under-use of potentially valuable is driving the government towards a sharing model for maximizing its use of its spectrum. As discussed above, much of the spectrum that might be allocated to mobile broadband is currently in the hands of federal government agencies whose use is often inefficient. Since moving these agencies is difficult and costly, spectrum sharing is an appealing policy.

2. Precedent for sharing government spectrum

Wi-Fi equipment operators have embedded low-cost radar detection chips in their devices – a primitive form of cognitive radio

Wi-Fi and radar systems sharing the 5 GHz band sets a strong precedent for government users sharing spectrum with private users. Wi-Fi equipment manufacturers have embedded low-cost radar detection chips in their devices – a primitive form of cognitive radio. Most industry observers cite this as a great success. But exporting this model to other frequencies may be more complex. There is an enormous difference in radiated power between Wi-Fi devices and radar that simplifies distinction. This asymmetry is not present in most other government uses so more complex systems are required to ensure compliance by lower priority (private) users. Still, dynamic spectrum allocation technology has rapidly developed in recent years (for more detail, see Section V.B below). Most industry observers believe that sharing is now technically viable in most frequency bands used by the federal government.

D. Broadcasters (Television and Radio)

Television broadcasters occupy 294 MHz of prime spectrum that many observers suggest is underutilized. Consequently, regulators and industry participants have focused significant attention on how to efficiently reallocate this spectrum.

1. Bands Used

The broadcast industry uses the following bands:

- **AM Radio:** 540 kHz to 1.7 MHz.
- **FM radio:** 92.1 to 107.9 MHz.
- **VHF Television:** 54 to 88 MHz and 174 to 216 MHz.
- **UHF Television:** 470 to 698 MHz.

2. Major Business and Regulatory Issues

The FCC issues television broadcasting licenses in 210 Television Market Areas (TMAs) for seven-year terms. Television broadcasting licenses are 6 MHz each. From a macroeconomic perspective, U.S. broadcast television is in a secular decline. Over

Television broadcasting licenses are 6 MHz each

90% of households now subscribe to satellite (DBS) or cable television rather than broadcast television.[50] Moreover, many economists believe that this spectrum would be better used if it were allocated to wireless data service providers. As a result, there has been a global movement to reallocate spectrum from television broadcasting to mobile data services.

The FCC has twice reduced the allocation for television broadcasting services over the past 30 years. In 1983, UHF channels 70 to 83 (806 MHz to 890 MHz) were re-allocated to the initial analog wireless services. In 2009 when the United States started the digital conversion of television broadcasters, channels 52-69 (698 MHz to 806 MHz) were "repacked" into lower channels. This opened the way for auctions of the 700 MHz band for mobile wireless. The FCC now seeks a third reduction of 120 MHz of the industry's 294 MHz in an incentive auction to eliminate another 20 additional channels.[51]

The FCC did not conduct auctions when it initially allocated television broadcasting spectrum

In an interesting twist, there is a small movement to allow low power television operators the authority to provide broadband service.[52] If the technical spectrum and hardware issues could be resolved, this might ease the conflict between television broadcasters and wireless broadband operators.

3. Thoughts on Valuation

The value of licenses is a function of the location of the license territory, the power designations, usage limitations, band location, terrain and strategic considerations (see Section VI for more information on valuation issues). The value of radio and television broadcasting licenses often differs significantly from one transaction

[50] NCTIA data and Summit Ridge Group, LLC analysis.

[51] The number of channels may be reduced by less than 20 (120 MHz reclaimed / 6 MHz per channel) if some existing stations agree to share 6 MHz license allocations. The FCC may also elect to reclaim less than 120 MHz.

[52] See: Broadcast Engineering, "Group Propose to Allow LPTV Stations to Deliver Wireless Broadband Service." June 2, 2011. http://broadcastengineering.com/rf/group-pushes-proposal-allow-lptv-stations-deliver-wireless-broadband-service.

to the next. The FCC did not conduct auctions when it initially allocated television broadcasting spectrum; the allocation cost broadcasters nothing. This further reduces publicly available pricing information. Moreover, commercial licenses tend to be sold only when a business is sold, making it difficult to unravel the value of the license from the going concern.

E. Satellite Communications Spectrum Dynamics

The satellite industry is focused on higher spectrum bands, but several of its lower bands have been targeted for reallocation.

C-band spectrum is underutilized in many regions.... Thus, many countries seek to reallocate the lower portion of this band for terrestrial or dual satellite /terrestrial use to ease the spectrum crunch

1. Bands Used:

- **L-band:** Satellite broadcasters are allocated the 1482.352 to 1490.624 MHz and 1525 to 1646.5 MHz bands. The former is used primarily for satellite radio and related services while the latter is for satellite telephony (MSS) services. The middle of the band is reserved for radio astronomy, military and satellite positioning systems such as the U.S. GPS, the European Galileo and the Russian GLONASS systems.
- **S-band:** 2.0 to 2.7 GHZ. This is used primarily for satellite radio services as well as mobile satellite services.
- **C-band:** 3.4 to 4.3 GHz (receive) and 4.25 to 6.425 GHz (transmit). Used primarily for satellite transmission and desirable in tropical areas due to its resistance to rain fade. Many countries are considering reallocating the lower portion of the receive band for terrestrial broadband, which is primarily used by Fixed Satellite Service (FSS) operators. A disadvantage of C-band is that it requires larger antennas on the ground.
- **Ku-band:** 12 to 18 GHz. Used primarily in satellite communications. Most commercial satellite television signals over the United States and Europe are transmitted in Ku-band. Although not as resistant to rain fade as C-band, its higher frequency allows consumers to use much smaller reception antennas. This band is used for DBS services (12.2 to 12.7 GHz) and FSS (12 to 14 GHz).

- **Ka-band:** 26.5 to 40 GHz. This satellite band was the newest addition to the commercial FSS bands.
  It has historically suffered from significant weather related interference, making it less valuable than C-band and Ku-band. But, higher-powered satellites developed over the past 10 to 15 years have largely overcome this problem. This band is heavily used in state-of-the-art satellite broadband systems. Local Multipoint Distribution Service (LMDS) is a terrestrial technology that operates in the Ka-band at 28 to 31 GHz to transmit signals from a single point to multiple points or a single point.

2. Major Business and Regulatory Issues

The satellite industry is seeing a trend of regulatory action to reallocate or share satellite spectrum to support terrestrial applications. Three major instances of this trend include:

The satellite industry is seeing a trend of regulatory action to reallocate or share satellite spectrum to support terrestrial applications

1) C-band spectrum is underutilized in many regions, particularly in areas not subject to heavy rainfall where C-band transmissions are largely unaffected. It is also increasingly feasible to use this band terrestrially due to the need for small cell sites in urban areas. Many countries are seeking to reallocate the lower portion of this band for terrestrial or dual satellite/terrestrial use to ease the spectrum crunch. Satellite operators, not surprisingly, are resisting this process.

2) The FCC recently began an Notice of Proposed Rulemaking ("NPRM") process to potentially allocate 500 MHz of Ku-band satellite spectrum in the 14 GHz band for two-way broadband service to airplanes via a terrestrial-based system.[53] While the FCC believes this can be done in a

---

[53] "In the Matter of Expanding Access to Broadband and Encouraging Innovation through Establishment of an Air-Ground Mobile Broadband Secondary Service for Passengers Aboard Aircraft in the 14.0-14.5 GHz Band" GN Docket No. 13-114 RM-11640, available at http://transition.fcc.gov/Daily_Releases/Daily_Business/2013/db0509/FCC-13-

manner that will not interfere with satellites, the satellite industry is highly concerned about the issue as the Ku-band is the most valuable satellite spectrum.

Mobile Satellite Services (MSS or "satellite telephony") operators seek increased rights to use their spectrum terrestrially

3) Mobile Satellite Services (MSS or "satellite telephony") operators seek increased rights to use their spectrum terrestrially. As the satellite telephony ventures of the 1990s largely failed, several of the license holders were able to lobby to have their spectrum rights expanded for terrestrial wireless service in conjunction with their satellite service. This was known as Alternative Terrestrial Component ("ATC"). Recently, DISH Network obtained relief from the ATC obligation and is now allowed to provide terrestrial service using phones without a satellite component. This obviates the expense of building a satellite network for dual satellite/terrestrial phones. The satellite industry is not opposing this action, presumably because it provides significant added value to the license holders.

3. Thoughts on Valuation

Orbital lots over oceans or less developed countries have much lower values, often for little more than the legal fees involved in securing the licenses

Most satellite spectrum is not auctioned as it falls under the ITU definition of international spectrum. Domestic spectrum is auctioned and licenses obtained though this process are often sold in an informal secondary market where prices vary considerably.

Satellite spectrum that targets large numbers of consumers, such as satellite television, is considered most valuable. In 1996, MCI bid a record $682 million in an FCC auction for the 110-degree West Longitude orbital slot for DBS that covered the entire continental United States. This record has not been surpassed. Satellite television appeared set for significant growth with only three slots covering the whole country. For comparison, the two satellite radio licenses sold in 1997 went for $83.4 million and $89.9 million, although they included far less spectrum.

66A1.pdf.

For FSS applications, spectrum value is much lower as most end users are businesses that can use multiple satellites with their reception equipment and can change to other satellites easily. However, locations over continents with prosperous end users have the highest values—up to $70 million.[54] Orbital slots over oceans or less developed countries have much lower values, often for little more than the legal fees involved in securing the licenses. Higher spectrum, such a Ka-band, sells for much less than C-band or Ku-band. However, orbital slot values are volatile and each situation must be carefully evaluated.

F. Government/Military Spectrum Dynamics

The U.S. government and military bands occupying approximately 60% of the spectrum below 3650 MHz are among the most valuable,[55] As described in Section III.C, the NTIA controls this spectrum.

1. But Government Efficiency is Questionable

[F]ederal government agencies do not have to compete for spectrum or justify their demand for it

The largest government users are the Department of Defense and the Federal Aviation Agency. As determined by the Government Accountability Office, federal government agencies do not have to compete for spectrum or justify their demand for the use of it. For this reason, federal spectrum users are under little pressure to increase efficiency.[56]

Although government use of spectrum has increased during the last few years, its inefficient use remains a serious issue in light of the shortage of frequencies available for commercial use. The rapid increase in the development and use of smartphones and tablets is driving greater demand for data-intensive services and spectrum. Spectrum shortage causes slow connections and dropped calls. Mobile service providers may have to react to the shortage by

---

[54] There are certain exceptions, most notably Eutelsat's and SES's "Hotbird" slots over Europe which would likely command far higher prices.

[55] PCAST Report, 8.

[56] PCAST Report, 55.

increasing prices and limiting the expansion of data intensive technological innovations.

2. It is Hard to Move Government Users

The FCC has allowed unused spectrum to be reallocated and used by unlicensed users under certain terms. But this reallocated spectrum came primarily from commercial entities rather than government users.

While it would be ideal for government users to vacate underutilized spectrum, this approach faces serious difficulties

Transferring under-utilized government-owned frequency bands to commercial entities would reduce or even eliminate the spectrum crunch as previous transfers have shown. When 90 MHz of spectrum owned by the government was reallocated to public entities, mobile 4G broadband services took its place.

While it would be ideal for inefficient government users to vacate underutilized spectrum, this approach faces serious difficulties. Much government spectrum, while rarely used, is reserved for emergencies. Moving or consolidating them would entail innumerable administrative hearings over the course of years. Additionally, many government agencies have custom equipment. The cost of replacing user terminals, antennas and repeaters, and conducting the radio frequency engineering needed to move to a new frequency is prohibitive. In the 1755 MHz band, the National Telecommunications and Information Administration (NTIA) concluded that clearing the 1755 to 1880 MHz band would cost $18 billion alone.[57] This would absorb most or all of the revenue that an auction might raise. Federal government investment in frequency bands below 3650 MHz totals approximately $281 billion. A breakdown of this investment is illustrated in Exhibit 4-6 below.

[57] NTIA,. "An Assessment of the Viability of Accommodating Wireless Broadband in the 1755-1780 MHz Band," 2012, available at www.ntia.doc.gov/report/2012/assessment-viability-accommodating-wireless-broadband-1755-1850-mhz-band.

Exhibit 4-6: Government Investment in Various Frequency Bands

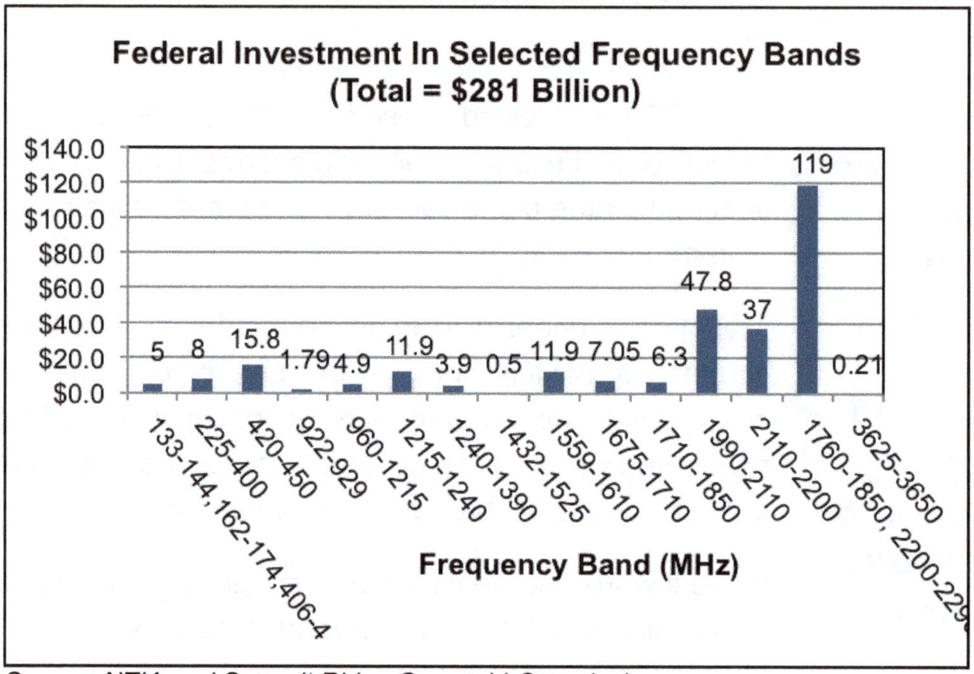

Source: NTIA and Summit Ridge Group, LLC analysis.

> ***[The PCAST report] concludes that sharing spectrum between existing government users and private users is a more efficient method [than clearing government users]***

As a result of the challenges in relocating inefficient government users, the PCAST report concludes that it does not believe that the FCC should clear and reallocate more government spectrum in the same way.[58] Instead, it concludes that encouraging sharing of spectrum between existing government users and private users is a more efficient method to increase utilization. Specifically, the PCAST report recommends that the government share 1,000 MHz of spectrum below 3.7 GHz with private users.[59] Pressure on government agencies to share spectrum is discussed in detail in Section V.C.

3. Cognitive Radios Enable Sharing

Cognitive radios can sense signals and utilize spectrum simultaneously on wide swaths of contiguous spectrum. However, there is some debate over the current state of technology for cognitive radios. With the database approach, radios check with a

[58] PCAST Report, 50.

[59] PCAST Report, 8.

central database (such as the ones maintained by Spectrum Bridge or Telecordia). The databases know the availability of the frequencies in different areas of the country. Cognitive radios can be designed to that must authorized by the database before transmitting. The database can assigned them an unused frequency based on their location. This may be an interim solution for Dynamic Spectrum Access until the viability of cognitive radio can be established.

Currently, much of the 5 GHz band is shared with aircraft radar. While this has generally been considered a success, it has not been without implementation challenges in the initial years of its rollout.

V. Sources and Timing of New Mobile Spectrum

The FCC has significantly increased the amount of mobile spectrum over the past several years, but not enough to meet projected demand growth. And forecasts suggest that the trend of increasing demand for mobile spectrum will continue. To address this challenge, the FCC has unveiled two strategies, which involve the reallocation of existing commercial spectrum from underutilized applications to mobile services and opening spectrum formerly reserved for government use for commercial users.

The 2010 National Broadband Plan is the centerpiece of the FCC's plan to expand spectrum for mobile broadband

A. National Broadband Plan and Television Incentive Auction

The 2010 National Broadband Plan is the centerpiece of the FCC's plan to expand spectrum for mobile broadband. This plan outlines a strategy to reallocate commercial spectrum from underutilized applications to mobile broadband.

1. FCC Seeks 300 MHz of Additional Mobile Wireless Specturm by 2015

By 2015, the FCC expects to add 300 MHz of frequency for mobile wireless services and another 200 MHz by 2020. The FCC seeks to achieve this, in part, by making 120 MHz spectrum in the 500 to 600 MHz range, currently used for broadcast television, available for mobile wireless, and by opening up frequencies as high as 3.6 GHz for the same purpose. See Exhibit 5-1 for the sources that the FCC has identified for additional mobile spectrum.

Exhibit 5-1: Sources of Additional Mobile Spectrum

Band	Key Actions and Timing	Megahertz Available for Mobile B-band	Current Status
WCS	2010 – Order	20	Assigned - 2012
AWS 2/3	2010 – Order 2011 – Auction	60	Pending
Upper 700 Mhz D Block	2010 – Order 2011 – Aution	10	Pending
Mobile Satellite Service (MSS)	2010 – L-band & Big LEO Orders; 2011 – S-band Order	90	Partially Assigned - 2012 (Globalstar)
Broadcast TV	2001 – Order 2012/13 – Auction 2015 – Band transfer/clearing	120	Pending
Total:		300	

Source: FCC: National Broadband Plan and Summit Ridge Group, LLC analysis.

The C-band is above the 3 Ghz range limit traditionally considered viable for mobile wireless service. But as technology improves and cell sites become closer, this spectrum becomes increasingly viable

2. FCC Seeks an Additional 200 MHz of Mobile Wireless Specturm by 2020

The FCC also seeks to add an additional 200 MHz of mobile broadband spectrum by 2020, in addition to the 300 MHz mentioned above. However, the FCC has yet to identify the source of this additional spectrum. Presumably it will reallocate under-used spectrum from other areas. One place the FCC might look is the lower portion of the satellite C-band in the 3.4 to 3.6 GHz range. Satellite operators have excess capacity in the spectrum band in most countries.

The C-Band is above the 3 GHz range limit traditionally considered viable for mobile wireless services. But, as technology improves and cell sites become closer, this spectrum becomes increasingly viable. Several governments have suggested that satellite operators share this spectrum with Wi-Fi users. The satellite industry is resisting this move. But senior satellite industry executives have confided to Summit Ridge Group that they consider the battle to maintain exclusive use of the lower portion of the C-band to be a "lost cause."

This timing of the satellite C-band spectrum move to wireless is uncertain. One option might be for spectrum to be moved to wireless broadband in stages. The first stage could be to share spectrum with the C-band for Wi-Fi. But Wi-Fi traffic and Wi-Fi offloading is rapidly rising.[60] If C-band satellites using lower C-band spectrum remain under-utilized, regulators might seek to reallocate it exclusively for Wi-Fi use.

Summit Ridge Group does not believe the FCC will meet its own deadline of allocating 300 MHz of spectrum to mobile broadband by 2015

3. FCC is Already Well Behind Schedule

Summit Ridge Group does not believe the FCC will meet its own deadline of allocating 300 MHz of spectrum to mobile broadband by 2015. The administrative processes for the elements listed in Exhibit 5-1 are already years behind schedule. We estimate that the most difficult reallocation – re-purposing 120 MHz of television broadcast spectrum -- will miss the deadline by three to five years.

One of the easier parts for the FCC to complete is authorizing the Mobile Satellite Service providers to use their spectrum terrestrially without the need for ATC authorization. As detailed in Section V.A.7 below, DISH received this authorization at the end of 2012. None of the other L-band or S-band satellite operators has yet received the authorization that the FCC planned for 2010 and 2011.

In fairness to the FCC, regulators in other countries are similarly behind schedule in reallocating spectrum. The European Union, for example, is well behind its goal of allocation an additional 1200 MHz for mobile broadband as outlined in its Radio Spectrum Policy Program (RSPP).[61] As described in Section VII, the U.S. has allocated more spectrum to mobile broadband than most developed countries and is at the forefront of allocating spectrum for shared use.

To reacquire 120 MHz, the government would need to acquire the equivalent of approximately twenty 6 MHz television broadcast licenses in each market

4. Reallocation of Television Broadcast Spectrum is Difficult

---

[60] Cisco. "Cisco Visual Networking Index: Global Mobile Data Traffic Forecast Update 2012-2017," 2.

[61] Available at: https://ec.europa.eu/digital-agenda/en/rspp-roadmap-wireless-europe.

The FCC issued a Notice of Rulemaking Proposal on October 2, 2012[62] indicating that it would conduct a "reverse incentive auction" for television broadcasters to sell their spectrum to the FCC. This would be followed by a "forward auction" in which mobile service providers would bid for the spectrum. The remaining broadcasters would be repacked to allow for the sale of spectrum in continuous blocks nationwide to broadcasters. The government has been careful to indicate that the auction process would be "voluntary," but the television broadcasters and the FCC have yet to agree on many details.

The incentive auction process has sparked tremendous resistance from the broadcasting industry, which sees it as threat to their ultimate spectrum rights and their own commercial viability. As previously mentioned, only 10% of television broadcast viewers receive their signals over-the-air. But the fact that the broadcasters broadcast signals over the air is what gives them the legal right to require cable companies to carry their signals. Between 35% and 40% of television broadcasters avail themselves to these so-called "must-carry" rules and would be loath to give them up. Broadcasters are also concerned about losing their spectrum through a later "involuntary" process should the FCC fail to meet its objective with its "voluntary" auction process. However, many broadcasters, particularly those without network affiliates, are struggling with declining revenue and see the incentive auction as a potentially attractive exit opportunity.

To reacquire 120 MHz, the government would need to acquire the equivalent of approximately twenty 6 MHz television broadcast licenses in each market. Most markets have significant unused spectrum so it may be possible. And as television stations would have the option of giving up a portion of their spectrum and "sharing" spectrum with another station, fewer than twenty stations would need to be taken, even in a full market. Summit Ridge Group

62 "In the Matter of Expanding the Economic and Innovation Opportunities of Spectrum Through Incentive Auctions; Docket No. 12-268,"September 2, 2012.

estimates that the FCC will only need to conduct these auctions in 30 to 40 markets to reacquire 120 MHz nationals since most markets have significant unused television broadcast spectrum.

5. Possible 600 MHz Band Plan

On October 2, 2012, the FCC released a notice of proposed rulemaking that included specifications for a possible band plan for the 600 MHz band after the auction process. It consists of an uplink that starts at 698 MHz and continues down, and a downlink that starts below Channel 37 (reserved for radio astronomy) and continues down. It is illustrated in Exhibit 5-2 below.

Exhibit 5-2: Post-Incentive Auction Band Plan

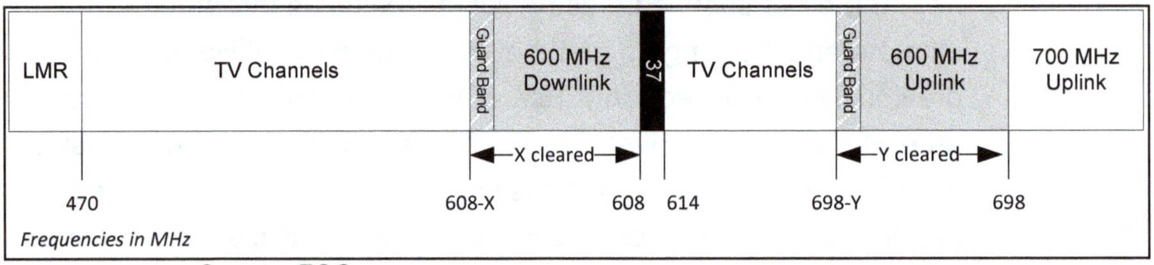

Source: FCC

Both television broadcasters and the wireless industry executives have criticized the FCC's proposed [600 MHz] band plan

Both television broadcasters and the wireless industry executives have criticized the FCC's proposed band plan. The biggest problem is the large space, called "duplex space," between the uplink and the downlink for the mobile service providers. Ideally, for LTE service, the gap would be no more than 4% to 5%, or about 30 MHz. The FCC's plan calls for approximately 130 MHz of duplex gap, assuming thirty 6 MHz television stations, plus channel 37 and buffer spectrum (generally referred to as "guard band" spectrum) between the bottom of the uplink and the top of the television channels.

Recently the NAB and the wireless industry submitted a counter-proposal to keep all of the wireless spectrum together at the top of the band to channel 30 (698 to 572 MHz) while putting the television spectrum at the bottom of the band (572 and lower). The wireless providers prefer this approach as it minimizes the duplex gap and facilitates future reconfiguration such as asymmetrically sized uplink and downlink blocks. The television broadcasters prefer the lower frequencies, which offer their channels better coverage.

A third option is to allocate the spectrum for LTE-TDD services. As described in Section IV.A.5.b, TDD does not use separate uplink and downlink bands so it may be better suited to asymmetrical broadband traffic. However, few U.S. mobile operators use TDD. The additional equipment costs to build out the spectrum TDD would likely depress auction revenue.

Parts of the 700 MHz band previously allocated for television channels 62-64 and 67-69 have also been allocated for a nationwide public safety network dubbed "FirstNet."[63] The U.S. Congress has authorized $7 billion for the FirstNet project, funded by the 600 MHz auction.

...ATC authority allows licensees to use satellite telephony spectrum to provide service terrestrially at the same time as using it to provide service via satellite

6. ATC Authorization increases value

MSS licensees may apply for an additional authorization to their licenses pursuant to the Ancillary Terrestrial Component (ATC) rules. These rules enable them to simultaneously use a frequency for both terrestrial and satellite uses. More specifically, ATC authority waivers allows licensees to use satellite telephony spectrum to provide service terrestrially at the same time as using it to provide service via satellite. To date, Globalstar, Inmarsat, Skyterra (now Lightsquared) and Terrestar have received ATC waivers.

Inmarsat PLC and LightSquared Co. (formerly Skyterra), both with ATC waivers, have L-band frequencies that are broken into interspersed pieces: Inmarsat/LightSquared/Inmarsat/LighSquared etc. To gain contiguous spectrum and maximize their capacity for 4G wireless services, LightSquared has agreed to a spectrum swap whereby it would pay Inmarsat in return for swapping its spectrum such that their respective spectrum becomes contiguous as opposed to interspersed. However, in February 2012, the FCC

[63] For information about FirstNet, see the National Telecommunications & Information Administration website at: http://www.ntia.doc.gov/category/firstnet.

withdrew its approval for a waiver of the ATC requirements as they apply to LightSquared. The FCC did this due to the potential for LightSquared signals to interfere with GPS receivers that frequently enter LightSquared's spectrum. Thus LightSquared cannot use its spectrum for ATC at this time. LightSquared subsequently entered Chapter 11. While there is industry speculation that the FCC will attempt to find LightSquared replacement spectrum, the fate of the company and its spectrum arrangement with Inmarsat remain unclear.[64]

The L- and S-bands are increasingly valuable frequency ranges due to the integration of satellite radio services and satellite mobile

The L- and S-bands are increasingly valuable frequency ranges due to the integration of satellite radio services and satellite mobile services. Along with other spectrum bands using ATCs, the L- and S-bands are capable of providing mobile services since the voice and data functions they provide are similar to those provided by terrestrial-based bands for cellular services. Another advantage of the L- and S-bands with ATC authorization is the increased coverage range compared to conventional terrestrial cellular networks. However, the ATC spectrum has significant disadvantages. The ATC rules for terrestrial spectrum use require service providers to integrate the network with a costly satellite platform thereby reducing its value. Also, the high cost of handsets needed for dual mode terrestrial/satellite-based services further increases the cost of the system.

Globalstar has petitioned the FCC for authorization to use its 2.4 GHz spectrum… for Wi-Fi service

7. DISH's AWS-4 Spectrum and Globalstar's 2.4G Petition

On December 11th, 2012, the FCC eliminated ATC requirements for DISH's AWS-4 spectrum between 2000 and 2020 MHz and 2180 and 2200 MHz. This includes the expensive requirement that all terrestrial handsets be integrated with the satellite network. The spectrum is now available for stand-alone terrestrial use as well as

---

[64] Outgoing FCC Chairman Julius Genachowski recently predicted the LightSquared would eventually get approval to operate, perhaps with a spectrum swap. See: Moritz, Scott and Todd Shields, "FCC Chief Sees LightSquared Getting Cleared for Airwaves Use." May 8, 2013. Bloomberg BusinessWeek, Available at: http://www.businessweek.com/news/2013-05-08/fcc-chief-sees-lightsquared-getting-cleared-for-airwaves-use .

MSS use. Summit Ridge Group believes that the FCC will show similar flexibility to the other ATC providers to encourage greater use of the spectrum.

Globalstar has petitioned the FCC for authorization to use its 2.4 GHz spectrum (2483.5 MHz to 2495 MHz) and the adjacent guard band (2473 MHz to 2483.5 MHz) for Wi-Fi service. Currently the Wi-Fi spectrum just below Globalstar's, 2473 MHz to 2483.5 MHz is largely unused because there is not enough spectrum to provide a full Wi-Fi channel without interfering with Globalstar's system. But if Globalstar can use both groups of frequencies, it can provide the market a full channel of additional Wi-Fi service that would otherwise be unavailable. The FCC will need to consider potential interference issues with the adjacent Wi-Fi channel. However, many industry observers expect the FCC to issue an NPRM on the issue shortly.

There is a total 90 MHz of broadband capable MSS spectrum (split between the L-band and the S-Band) including DISH's AWS-4 spectrum as shown in Exhibit 5-3 below:

Exhibit 5-3: Broadband and Broadband Capable MSS Bands

MSS Band	Allocated Bandwidth	Bandwidth Usable for Terrestrial Broadband	Licensees	Subscribers
L-band	Two 34-megahertz blocks at 1525–1559 MHz, 1626.5–1660.5 MHz	40 megahertz	SkyTerra	18,235
			Inmarsat	254,000
S-band[65]	Two 20-megahertz blocks at 2000–2020 MHz, 2180–2200 MHz	40 megahertz	DBSD (ICO)	-
			TerreStar	-
Big LEO	Two 16.5-megahertz blocks at 1610–1626.5 MHz, 2483.5–2500 MHz.	10 megahertz	Globalstar	382,313
			Iridium	359,000

Source: FCC National Broadband Plan and Summit Ridge Group, LLC analysis.

Other ideas for spectrum have been suggested… one challenge common to each is that, in most cases, there is insufficient hardware available to use the spectrum in the short-run

As a result, ATC could be a significant addition to spectrum capacity should the FCC continue its precedent with the aforementioned DISH waiver and allow a seamless addition to the supply of terrestrial mobile wireless spectrum without the expense of building a satellite network.

8. AWS-1, AWS-2 and AWS-3 Potential

AWS-2 spectrum contains two blocks: the H Block including frequency ranges 1915 to 1920 KHz and 1995 to 2000 MHz, and the J Bock including frequency ranges 2020 to 2025 MH, and 2175 to 2180 MHz. Proposals for allocating this spectrum are pending before the FCC. The FCC's 2004 Notice of Proposed Rulemaking for this spectrum is also still pending. The H block is likely to be most valuable to Sprint which owns the adjacent G Block.

The AWS-3 spectrum runs from 2155 to 2175 MHz and is not paired with another frequency block. Proposals for this spectrum, including pairing it with the 1755 to 1780 MHz band, are currently

---

[65] *This spectrum, controlled by DISH, was authorized to for terrestrial mobile use and is referred to as AWS-4 spectrum discussed in Section V.A.7 above.*

pending before the FCC. The FCC is expected to allocate both the 1755 to 1780 MHz band and the 2155 to 2180 bands by 2015.

AWS is primarily used for cellular 3G and, to a lesser extent, 4G services. One advantage of this spectrum is its suitability for a range of uses due to its solid propagation capacity (although it has a smaller coverage range than lower frequencies such as the 700 MHz band). For the spectrum to be utilized efficiently, two issues must be resolved. First, existing government users in the 1710 to 1755 MHz range (Uplink for AWS-1) must relocate (see Exhibit 9-7 for AWS-1 band plan). Second, more equipment must be developed to support AWS. However equipment manufacturers are reluctant to invest in new technologies given the uncertainty of the timeframe for the band clearing.

B. 65 MHz in Required Auctions per Spectrum Act

Section 6401 of the "Middle Class Tax Relief and Job Creation Act of 2012," known as the Spectrum Act, directs the FCC to auction several spectrum blocks, totally 65 MHz, by early 2015. These blocks are shown in Exhibit 5-4 below.

Exhibit 5-4: 65 MHz of Spectrum to be Auctioned by 2015

Band	Amount	Comments
1675 - 1710 MHz	15 MHz	FCC to identify 15 MHz within the 35 MHz block to auction
1915 – 1920 MHz	5 MHz	Lower AWS H Block – likely to be paired with below
1995 – 2000 MHz	5 MHz	Lower AWS H Block – Likely to be paired with above
2155 - 2180 MHz	25 MHz	AWS-3 Band with upper AWS J Bock
???	15 MHz	FCC to auction 15 MHz in spectrum it determines
Total	**65 MHz**	

The addition of 65 MHz of mobile broadband spectrum is

significant. However, it only increases the current allocation of 608 MHz by just over 10%. Meanwhile, mobile broadband usage, by any measure, is growing much more rapidly.

C. Other Sources of Additional Mobile Spectrum Capacity

Several other ideas for additional spectrum have been suggested. This section addresses those that are discussed most. A challenge common to each of them is that, in most cases, there is insufficient hardware available to take advantage of the spectrum in the short-run. While the development of smart radio technology may ameliorate this challenge, we are years away from wide-scale availability at competitive prices.

1. Wi-Fi and WiMax

Broadband Radio Service (BRS), formerly allocated for wireless television as MMDS band, is a spectrum range between 2.5 and 2.7 GHz that is primarily used for two-way integrated communication services such as voice, data and Internet services transmitted through cellular systems. The BRS spectrum currently offers an opportunity for significant expansion of services transmitted in the WiMax format. This is because, currently, only a fraction of the 194 MHz of spectrum in the 2.5 GHz band has been built out. The WCS band also includes WiMax-based and the FCC recently announced a plan for opening additional spectrum in the 5 GHz band for future Wi-Fi services.

Closer cell spacing mitigates the disadvantage of the distance limitations of 2.5 GHz spectrum

Until recently BRS was much less valuable than lower frequencies since much of it remained unused despite the availability of commercial equipment. This was due to its inferior building penetration and limited distance range. But, as spectrum has become more crowded, wireless companies have placed cell sites closer together, particularly in metropolitan areas. Closer cell spacing mitigates the disadvantage of the distance limitations of 2.5 GHz spectrum. While closer cell spacing is expensive, it also increases system capacity by increasing the number of times frequency is reused, which allows greater power levels needed for higher data transmission rates). At the same time, recent equipment developments have reduced building penetration obstacles. Consequently, this spectrum has become more desirable for mobile wireless applications, particularly in heavily used urban markets.

The FCC plan to use the 2.5 GHz band to expand wireless spectrum capacity is shared by other countries around the world. The ITU has created three recommendations for dividing the band as described in Exhibit 5-4 below. Several countries in the Americas including Brazil, Canada, Chile and Columbia have adopted Option 1. The U.S. is currently not in compliance with any of these options, but, one party, Clearwire, controls most of it. Should equipment manufacturers and market forces effect harmonization with other countries in the Americas, it should be relatively easy for them to accomplish the FCC's goal of using the spectrum for additional wireless spectrum capacity.

Exhibit 5-4: ITU Recommendations for the 2.5 GHz Band

Frequency arrangement	Mobile station transmitter (MHz)	Centre gap (MHz)	Base station transmitter (MHz)	Duplex separation (MHz)	Centre gap usage
C1	2 500-2 570	50	2 620-2 690	120	TDD
C2	2 500-2 570	50	2 620-2 690	120	FDD DL (external)
C3	Flexible FDD/TDD				

Source: ITU Recommendation M1036-3[66] *and Summit Ridge Group, LLC analysis.*

The FCC has often been reluctant to enforce its rules related to build out milestones

2. More Aggressive FCC Rule Enforcement

The FCC has often been reluctant to enforce its rules related to build out milestones. It likely fears driving away smaller new entrants, who are often the most promising innovators. But these smaller entrants are often capital constrained and encounter delays in being able to finance the required build outs. It has been Summit Ridge Group's general observation that the FCC is frequently flexible in granting waivers for various construction delays,

[66] Available at: http://www.itu.int/rec/R-REC-M.1036-3-200707-S/en

particularly with new entrants.

Providing waivers for required build out milestones may help new undercapitalized entrants. But it also leaves valuable spectrum underutilized. A more aggressive enforcement of FCC license rules could provide additional spectrum. However, the FCC needs to balance those benefits against the possible long-term costs of fewer new entrants if stricter license rules deter investors.

3. Un-auctioned or Reclaimed Federal Spectrum

The FCC has resisted selling [spectrum that it failed to sell at auction] on the secondary market

In many previous auctions, the government failed to sell the entire spectrum available for sale. Summit Ridge Group's analysis of Scott Wallsten's FCC auction data[67] shows that of the 81 FCC Spectrum Auctions that have taken place since 1994, the FCC failed to sell all of the licenses in 37 of them. Additionally many purchasers returned or forfeited their spectrum rights because they were unable to meet various build-out requirements. As a result, the FCC holds numerous slivers of spectrum that remain unallocated. The FCC has resisted selling this spectrum on the secondary market. Rather, they seek to hold the spectrum until there is enough to justify an additional auction. In certain bands, this is not likely to happen for several years.

4. Opening Unused Spectrum for Sharing

Another concept to increase spectrum availability is to open unused licensed spectrum, on a temporary basis.[68] Smart radio technologies arguably make this idea increasingly viable. Licensees of the unused spectrum may oppose this out of fear that once people start using the spectrum in large numbers, it may be difficult to remove them. There is also a risk of a "tragedy of the commons" whereby spectrum use is dominated by low-priority uses because it is available at no cost, crowding out more valuable uses. One

[67] Scott Wallsten, "Is there Really a Spectrum Crisis? Quantifying the Factors Affecting Spectrum License Value." Technology Policy Institute, 9.

[68] For a detailed analysis of this proposal, see Michael Calabrese, "Use it or Share it: Unlocking the Vast Wasteland of Fallow Spectrum," September 25, 2011. TPRC 2011, available at SSRN: http://ssrn.com/abstract=1992421.

solution would be to have micro-payments in crowded markets.[69] One small mobile operator, Republic Wireless, offers a service for $19 a month that uses unlicensed Wi-Fi spectrum and only "roams" to licensed spectrum if unlicensed spectrum is unavailable.

The invention of spectrum sensing radios and spectrum databases has made it possible to share spectrum dynamically

D. Government Spectrum Sharing

1. Smart Radios and Spectrum Databases Enable Sharing

The invention of spectrum sensing radios and spectrum databases has made it possible to share spectrum dynamically. Spectrum sensing radios detect unused spectrum while spectrum databases monitor allocated and unallocated spectrum as well as the priority of each user. A spectrum sensing radio would detect an unused channel, then verify with the database to see if the spectrum is reserved for a higher priority use before allowing the radio to operate in that spectrum. If a higher priority user signaled that they wanted to use the spectrum, the database would shut down or move the lower priority user to available spectrum. This process is illustrated in Exhibit 5-5 below.

[69] For an analysis of this proposal, see Eli Noam, "The Next Stage in the Digital Economy: Nano-Transactions and Nano-Regulation" Columbia Institute for Tele-Information. December 2000, available at: http://www.citi.columbia.edu/elinoam/articles/con_info_money.htm

Exhibit 5-5: Sharing Mechanics

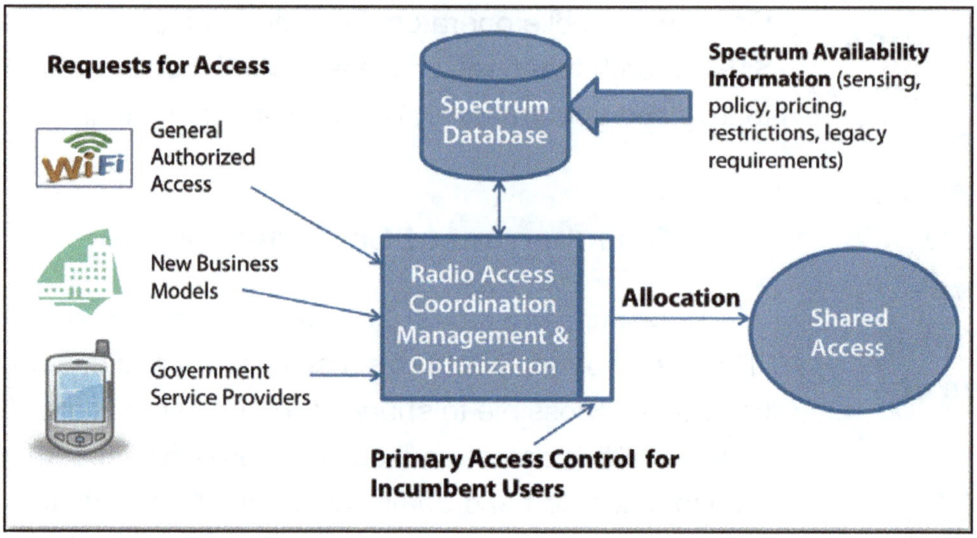

Source: PCAST Report.

The PCAST Report recommends improving spectral efficiency by requiring federal agencies to "share" unused spectrum with commercial users

Current spectrum-sharing research proposals include the Authorized Shared Access (ASA)/Licensed Shared Access (LSA) system, the Wireless Access Policy for Electronic Communications Services (WAPECS). The LAS and WAPECS are European Union policy concepts. Other spectrum sharing proposals include those developed from the white space trials carried out in the United Kingdom.

2. Government Users Pushed to Share Spectrum

As mentioned in Section IV.C.2 above, the PCAST report recommends against moving government users. Instead it proposes improving spectral efficiency by requiring federal agencies to "share" unused spectrum with commercial users. This would build on the success of Wi-Fi sharing radar spectrum in the 5 GHz band.

Due the availability of modern spectrum sensing radios, such sharing is potentially viable from a technical perspective. A chart showing federal spectrum under consideration for sharing is shown in Exhibit 5-6 below. The PCAST report also recommends the sharing of commercial spectrum when not used by the primary licensee.

Exhibit 5-6: Federal Spectrum Under Investigation for Shared Use

Frequency Band (MHz)	Amount (MHz)	Current allocations/usage (Federal, non-Federal, Shared)
406.1 - 420.0[1]	13.9	Federal
1300 – 1390[1]	90	Federal
1675 – 1710[2]	35	Federal/non-federal
1755 – 1780[2]	25	Federal
1780 – 1850	70	Federal
2200 – 2290	90	Federal
2700 – 2900[1]	200	Federal
2900 - 3100	200	Federal/non-Federal shared
3100 – 3500	400	Federal/non-Federal shared
3500 – 3650[2]	150	Federal
4200 - 4400[1] [4200-4220 & 4380-4400]	200	Federal/non-Federal shared Federal/non-Federal shared
Total	**1,473.9**	

[1] Band obligated by U.S.-Canada or U.S. Mexico bilateral agreement(s).
[2] Bands selected for Fast-Track Evaluation. For purposes of future analysis, 1755-1850 MHz – consisting of 1755-1780 MHz and 1780-1850 MHz – will be assessed as a single block.

Source: PCAST Report and Summit Ridge Group, LLC analysis.

The PCAST Report suggests a three-tier system of rights for sharing government spectrum. Under this plan, government users would have first priority to their spectrum; second priority would be to commercial users who would pay a fee for using spectrum on a temporary basis. The lowest priority would go to users who could use any remaining unused spectrum at no charge. An illustration of this hierarchy is shown in Exhibit 5-7 below.

Exhibit 5-7: Suggested Sharing Hierarchy of Government Spectrum

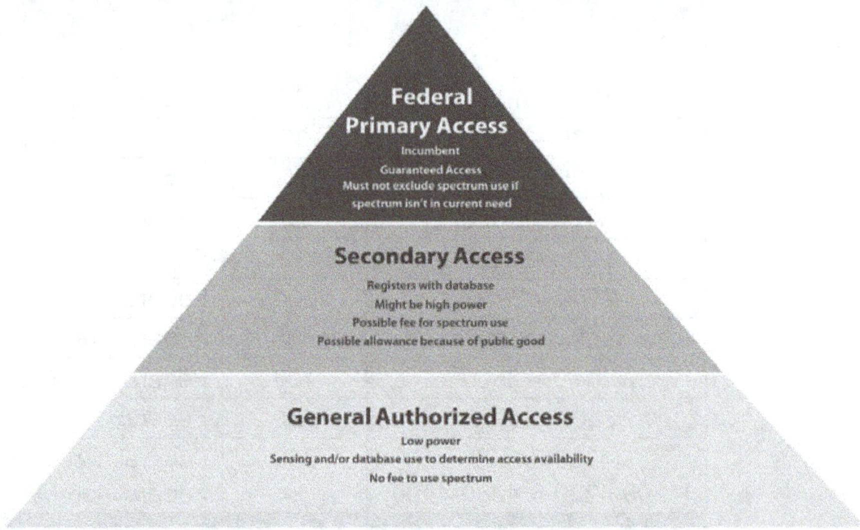

Source: PCAST Report.

A major implication resulting from spectrum sharing is the possibility of creating intermediate- and short-term spectrum licenses

3. Greater Licensing Flexibility Possible

A major implication resulting from the development of spectrum sharing is the possibility of creating intermediate- and short-term spectrum licenses. Under the current regulatory system, a long-term spectrum license is available at auction or in the secondary market. Unlicensed spectrum can also be used, but with no guarantee of future availability or freedom from interference.

Under a spectrum sharing scenario, second tier users could reserve spectrum for short or intermediate-terms by the hour, day, week or month. The primary users can grant access for such periods with assurances of availability through the database authorization system. This treats spectrum like other property that can be secured for various periods. This is shown in Exhibit 5-8 below.

Exhibit 5-8: Potential for Increased License Flexibility

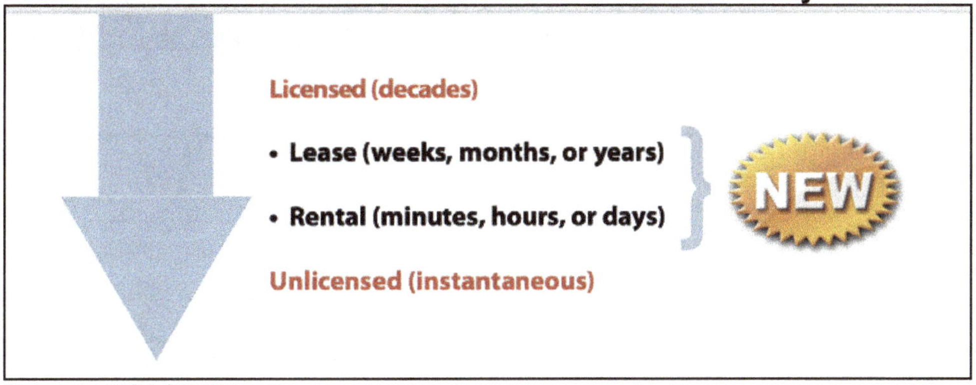

Source: PCAST Report.

4. Timing of Federal Sharing

Summit Ridge Group believes shared government spectrum allocation will grow over time as the enabling technology hurdles and business models are settled

The NTIA appears to be taking the push to share federal spectrum seriously. It is currently working on a "Test Bed Pilot Program" to evaluate the possibility of dynamic spectrum access. Dynamic spectrum involves using technology such as smart radios and spectrum databases that allow spectrum users to change frequencies as needed on a continuous basis. Such technology would be critical for commercial users to be able to share sensitive federal spectrum on a non-interfering basis with rules potentially modeled after the TV white spaces in commercial spectrum.

Spectrum sharing is being evaluated and/or implemented by several developed countries. The international nature of the movement gives it significant credibility and momentum. Summit Ridge Group believes shared government spectrum allocation will grow over time as the enabling technology hurdles and business models are settled. However, given the significant nature of the policy change it will be years before government agencies fully embrace this option.

As shown in Exhibit 5-9, there is nearly 500 MHz of shared spectrum allocation currently in the FCC NPRM process or at other levels of advanced consideration.

Exhibit 5-9: Shared Spectrum in the FCC NPRM Process

Band	Amount	Status
1695 to 1710 MHz	15 MHz	
1755 to 1850 MHz	95 MHz	NTIA analysis
3550 to 3650 MHz	100 to 150 MHz	GN Docket No. 12-354
4940 to 4990 MHz	50 MHz	WP Docket No. 07-100
5.35 to 5.47 GHz, 5.85 to 5.925 GHz	195 MHz	ET Docket No. 13-49
Total	**440 to 490 MHz**	

Source: FCC and Summit Ridge Group, LLC analysis

Potentially, some or all of this spectrum could become available over the next year or so. This could considerably change the spectrum supply dynamics.

VI. Spectrum Valuation Analysis

Spectrum has three dimensions of value: 1) the financial benefits for the licensee; 2) the indirect economic benefits to society; and 3) the social benefits such as improved access to emergency services. This section will focus only on the first dimension of value, while acknowledging the other values may be much higher. A complete analysis, even of the financial value to the licensee, is beyond the scope of this paper. Rather, this section will provide an introduction to the process. Summit Ridge Group has extensive experience in spectrum valuation and is available for more detailed consultations.

Historically, spectrum price volatility, particularly for less desirable spectrum, has been very high

Summit Ridge Group believes that strict regulation and rapid demand growth has made spectrum a scarce resource in the United States and other parts of the world.[70] Historically, spectrum price volatility, particularly with less desirable spectrum, has been very high. When estimates for demand increase, the value of less desirable spectrum spikes on the expectation that well-financed service providers will soon develop it. As demand estimates fluctuate downward, prices plummet on doubts that it will ever be put to profitable use. However, there are analytical frameworks for considering spectrum value.

Spectrum licenses are intangible assets, which technically cannot be considered property since the licensee must explicitly waive ownership claims over the acquired spectrum. For valuation purposes, spectrum licenses can be treated as quasi-property. There are three approaches to consider when evaluating the commercial value of spectrum – the income, market and cost

---

[70] The debate over the role government regulation should play in the process is far from settled. In general, European policy makers favor a larger and more deliberate government role in spectrum management than its counterparts in the U.S. Spectrum is, in theory, infinite since cell sizes can be made smaller and smaller. However, such infrastructure build out becomes cost-prohibitive.

approaches. These approaches seek to answer the following questions: How much value does it add? What is the market price? and What would it cost to replace? Additionally, policymakers often use econometric modeling to gauge general valuation levels for auction estimation or tax purposes.

Industry custom is to describe spectrum pricing in units of price per MHz/PoP, i.e. the price divided by the number of megahertz, divided by the population covered. While not all MHz or populations are equal, this calculation provides a consistent comparison of prices across transactions.

A. Income Approach

1. Description

The income method predicts future benefit streams (usually cash flows) that a licensee might generate using the acquired spectrum. The predicted cash flow in excess of the costs required to generate revenue (including the cost of spectrum, equipment, labor and the normalized cost of capital) are discounted for risk and the time value of money. The result is the intrinsic value of the spectrum license. The most common version of the income approach is called the discounted cash flow (DCF) analysis.

2. Application

The income method is most useful when used internally since accurate and reliable estimations of future revenues and growth are needed

The income method is most useful when used internally since accurate and reliable estimations of future revenues and growth are needed. This is often the case with stable industries.

For licenses with established uses such as radio broadcasting, industry metrics are used to estimate the potential value of a license – a critical element for starting the business. This is called the Greenfield Method, which is essentially a DCF analysis starting from scratch. Another version of the income method is called the Multi Period Excess Earnings Model (MPEEM). The MPEEM can be used to value a critical business intangible. It examines the return a business earns and subtracts the portions attributable to other supporting assets. The MPEEM allocates the remainder to the license (or other critical asset) and discounts it. A complete description of the MPEEM models is beyond the scope of this text,[71] but Summit Ridge Group professionals have experience in applying it and explaining the methodology to clients and in legal proceedings.

3. Challenges

The income method is very sensitive to the projections and discount rates. Small changes in assumptions can result in large differences in valuation. With early stage companies, particularly those with new technologies or business models, long-term projections are typically somewhat speculative.

B. Market Approach - Comparable Transactions

1. Description

The second method is based on historical data rather than predicted future data. It is called the market comparable method and is based on the logic that an asset is worth what has been previously paid for similar assets. It considers the price paid for similar licenses. A valuation is determined by applying the relevant metrics from prior transactions to the spectrum license being valued. Most intangible assets are unique so relevant market comparables are unavailable. Spectrum, however, is one of the few intangible assets where relevant market comparables are often available.

Spectrum is one of the few intangible assets where relevant market comparables are often available

2. Application

While the pricing standard is generally price per MHz/PoP, for satellite orbit slots and secondary licenses, different pricing standards are utilized. They are usually considered in terms of the price paid for other licenses in the same band. When using comparable transactions, industry-specific regulatory and technical issues must be considered and adjusted, which is a significant

---

[71] Additional information about the MPEEM can be found at: http://www.kpmg.com/CN/en/IssuesAndInsights/ArticlesPublications/Newsletters/Defining-Issues/Documents/Defining-Issues-O-1008-33.pdf

challenge. The market comparable method is the primary method used by audiences like courts or public market investors with limited industry data to scrutinize the assumptions in a DCF. The market approach is usually the best option if relevant comparables are available and cash flow uncertainty makes income-based approaches difficult.

3. Challenges

Direct comparables are often rare so adjustments are usually made to the closest available comparables. But it is often difficult to find principled methods for making the adjustments. When comparing auction results between countries, for example, seemingly small differences can have a large difference in expected auction outcomes, making adjustments challenging. Additionally, the market changes rapidly, particularly for non-prime spectrum, limiting the value of comparables.

Under the cost approach technique, the value of spectrum is the difference in costs between current or planned spectrum and the next best alternative

C. Cost Approach - Replacement Method

1. Description

The third method is based on the principle that something is worth no more than what it costs to replace it with something of equivalent functionality. For example, if a mobile operator loses spectrum, it can, for a price, replace that capability with additional cell splitting. This method, called the Optimal Deprived Valuation Method (or the "With and Without Method"), estimates the loss of value to the company if the spectrum were taken away and replaced by the most economic means for replicating the functionality of this spectrum using alternate means or otherwise adjusting to the loss in the most efficient manner.

2. Application

The most obvious way for a mobile operator to adapt to losing spectrum is by further dividing existing cell sites in the remaining spectrum into smaller cells to increase reuse. The cost of this provides an approximate value of the spectrum. Exhibit 6-1 below shows the FCC projections for the cost savings to the wireless industry of providing 500 MHz of additional spectrum.[72]

Exhibit 6-1: Cost Savings of Additional Spectrum

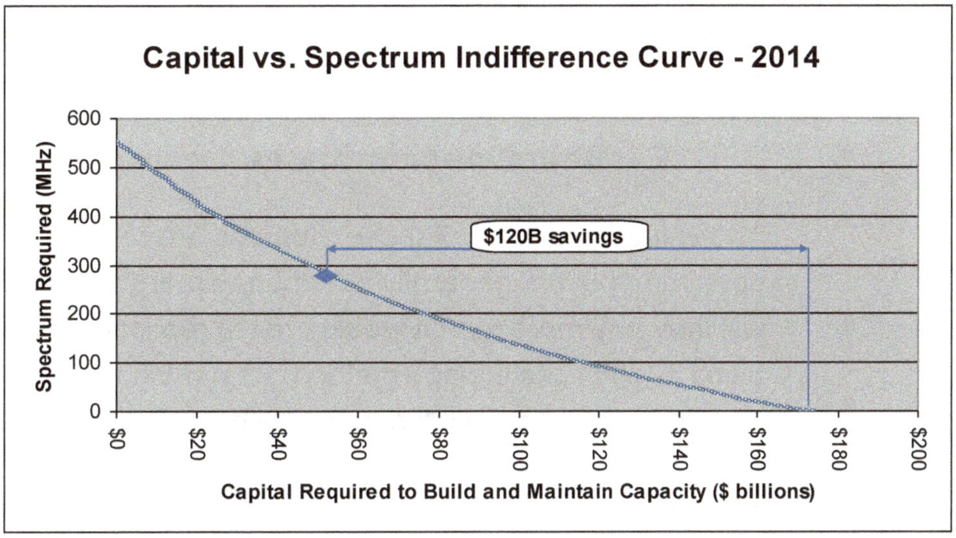

Source: FCC – "Mobile Broadband: The Benefits of Additional Spectrum."

Regulators often use econometric modeling to estimate auction proceeds or to set tax levels. Business valuation professionals evaluating a specific asset generally do not use it

Using this technique, the value of spectrum is the difference in costs to a company between its current or planned spectrum and the next best alternative. The Optimal Deprived Valuation Method is most applicable for developed businesses trying to sell spectrum for which there is a clear alternative replacement at a known cost.

3. Challenges

This method is not a viable method for businesses that would need to shut down when faced with spectrum reduction. It is also often difficult to predict the most efficient manner for a company to respond to the loss of an asset, or to predict the replacement cost of a comparable asset.

D. Econometric Modeling

---

[72] For an additional analysis of the trade-off between additional spectrum and capital expenditures, see Robert J. Shapiro, Douglas Holtz-Eakin, & Coleman Bazelon, "The Economic Implications of Restricting Spectrum Purchases in the Incentive Auctions," April 30, 2013, available at: http://www.gcbpp.org/files/Academic_Papers/SpectrumAuctions.pdf.

This method analyzes statistical differences between variables and a sample of other assets. It is useful when determining a general value for a class of assets than for a specific asset. Regulators often use it to estimate aggregate auction proceeds or to set tax levels. Business valuation professionals evaluating a specific asset generally do not use it.

E. Other Considerations

The valuation drivers of these methods are related. When valuing spectrum, buyers generally will not pay more the market price, nor will they pay more than it would cost to replicate the functionality using another technology method. Assuming those conditions are met, their valuation will focus on the potential return to a business using the spectrum. This return depends on many factors: the network construction cost, the quantity of customers that the spectrum can support, FCC regulation, topography, availability of equipment, required battery life, etc.

Service providers generally will not pay more than the market price for spectrum, nor will they pay more than the cost of replicating the functionality using another technology method

The cost of building a network largely determines the economic feasibility of adopting a particular frequency. The cost of using a particular frequency is, in turn, determined by the availability of equipment developed for that spectrum, the supply of towers and the spectrum's cost. The cost of spectrum varies according to its physical propagation characteristics in the region, the costs to build a network using it as well as strategic issues such as the equipment vendor support needed to make a spectrum range valuable for commercial use. With more equipment vendor support, more equipment becomes available, and the particular frequency becomes cheaper and easier for operators to build out.

The number of customers that can be supported by the frequency is also important. The more customers it can support, the more valuable is the frequency. This is partially dependent on the propagation qualities of frequency and the local geography in terms of the needs of the buyer.

F. Valuation Challenges

1. Business Environment Can Change

Industries are dynamic. The numbers of players change, significantly influencing profitability. Industry dynamics can also impact the number of potential buyers, affecting the liquidity of the market in which a license is sold. There is often uncertainty about the future demand for services on the spectrum. Strategic issues are also significant, particularly with new LTE technologies for which operators prefer large (15 or 20 MHz) swaths of contiguous spectrum. A license holder with spectrum in the middle of a larger holder's otherwise contiguous swath may have significant economic leverage.

Demand growth can also change significantly. For example, the demand for wireless broadband is currently growing rapidly. Wireless operators are investing in new technology to accommodate this growth. However, should growth slow markedly, reverse, or be absorbed by unlicensed spectrum, the value of wireless broadband spectrum could dramatically decline, particularly for spectrum towards the higher or lower ends or traditional use for a given application.

Wireless operators normally prefer licensed spectrum to shared spectrum

2. Technological Change Risk

The communications industry is technology industry, which, by nature, is constantly in flux. Change increases or decreases the value of spectrum depending on the situation. Currently, there is a perceived shortage of spectrum, which increases pricing as new technologies encourage wireless users to use more data intensive applications. Other new technologies such as dynamic spectrum access radios and associated databases promise to significantly increase the supply of spectrum through sharing, depressing spectrum valuation. Wireless operators normally prefer licensed spectrum to shared spectrum because the availability of licensed spectrum is guaranteed for their customers. But, the alternative of unlicensed spectrum and technical improvements in cognitive radio technology may limit the price operators will pay for licensed spectrum.

3. Financial Market Risk

Markets change rapidly – even more rapidly than technology. This impacts the theoretical value of spectrum as the cost of capital changes. It also affects a buyer's access to capital as financing windows open and close. Typically, the cost of capital and access to capital markets improve or deteriorate at roughly the same time, thus compounding the challenge.

4. Regulatory Risk

The telecommunications industry is highly regulated, particularly with respect to spectrum. Regulatory authorities increase the supply of spectrum thus lowering demand for spectrum at a given price-point. They also can change license terms, enforcement terms, and grant or deny waivers for terms such as build-out requirements, and control many other factors that impact value. Regulatory changes are often difficult to predict, both in terms of the substance and the timing of those changes. Exhibit 6-2 shows general ranges for the impact of common regulatory factors.

Regulatory changes are often difficult to predict, both in terms of substance and the timing of those changes

Exhibit 6-2: Impact of Common Regulatory Factors

Driver	From (worst)	To (best)	Impact on cost to serve subscribers
Infrastructure/ Spectrum Sharing	No site/spectrum sharing allowed	Active site and spectrum sharing encouraged	-35% to -40%
Coverage obligations	High coverage requirements	No coverage requirements	-25% to -35%
Industry structure	Multiple players	Economically viable number of players	-20% to -30%
Spectrum license fees	High fees	No fees	-15% to -30%

Source: ITU[73] and Summit Ridge Group, LLC analysis

5. Internal Strategic Issues

Spectrum can have value specific to the bidder due to its individual strategic issues. Spectrum may have value in blocking a competitor

[73] "Exploring the Value and Economic Valuation of Spectrum," International Telecommunications Union – Broadband Series, April 2013, 12.

in combination with existing spectrum, avoiding a site build-out, tax considerations and many other complex issues.

G. General Valuation Observations

Spectrum above 2 GHz or below 300 MHz has less value due to its difficulty with range and building penetration issues. However, as cell sites get smaller and new technology improves building penetration, spectrum near or above 2 GHz becomes more valuable

Many factors influence the value of spectrum. Detailing all of them is outside the scope of this report. But a few major considerations follow.[74]

1. Frequency Band is Critical

For terrestrial mobile wireless use, the most valuable spectrum is between 300 MHz and 1 GHz range since lower frequencies have better penetration qualities and travel further. However, spectrum between 1 GHz and 2 GHz has become valuable as well. Lower frequencies are in greatest demand for rural areas due to the larger circumference of area reached by waves. But in urban areas higher frequencies are often sufficient for high population density because the waves need not travel great distances.

Not only is the lower circumference of area reached by higher frequency waves no longer a disadvantage, but their improved data capacity becomes an advantage in urban areas. High frequencies enable spectral efficiency to increase for high-volume data applications. Since the demand for data-intensive services used on devices like smartphones and tablets has rapidly increased, the value of spectrum at the upper end of the spectrum rage has also increased.

Spectrum above 2 GHz or below 300 MHz has less value due to its difficulty with range and building penetration issues. However, as cell sites get smaller and newer technology improves building penetration, spectrum near or above 2 GHz becomes more

---

[74] For additional information on spectrum valuation see: Bazelon, Coleman and McHenry, Giulia, Spectrum Value (March 31, 2012). 2012 TRPC. Available at SSRN: http://ssrn.com/abstract=2032213 or http://dx.doi.org/10.2139/ssrn.2032213

valuable, particularly in urban areas where cell sites are the smallest. There is an ongoing movement to redeploy some of the satellite spectrum in the lower portion of the C-band (3.6 GHz) to Wi-Fi and other terrestrial wireless applications. A description of the coverage of various frequencies is in Exhibit 6-3 below.

The area covered by a cell is proportional to the square of its frequency

Exhibit 6-3: Coverage Difference Between Frequencies

Frequency (MHz)	Cell radius (km)	Cell area (km2)	Relative Cell Count
450	48.9	7521	1
950	26.9	2269	3.3
1800	14.0	618	12.2
2100	12.0	449	16.2

Source: Wikipedia and Summit Ridge Group, LLC analysis.

As the table above indicates, the area covered by a cell is proportional to the square of the frequency. Thus, at 2 GHz, four time as many cell sites are needed as at 1 GHz. The above table assumes the antenna size scales with wavelength. This is not always the case. If the antenna size is fixed, operators will experience less geographical coverage loss as one moves to higher frequencies. Moreover, other factors such as geography can also limit coverage and thus limit the ability of larger cells to utilize their theoretical full coverage areas.

2. Wider Bands of Paired Contiguous Spectrum Are More Valuable

To be efficient, newer spectrum protocols, including various spread spectrum technologies, require broader swaths of spectrum. Without a broad swath, spectrum is generally less valuable.[75]

a. Blocks of Contiguous Spectrum Valuable

Broader swaths of spectrum allow inter-cell distances to be larger

[75] Thomas W. Hazlett & Roberto E. Munoz, "What Really Matters in Spectrum Allocation Design," April 2010, available at http://businessinnovation.berkeley.edu/Mobile_Impact/Hazlett-Munoz_Spectrum_Matters.pdfhttp.

and improve efficiencies in the carriage of signal from network node nearest the origin to the node nearest the destination (also known as "trunking"). While 3G spectrum reaches its maximum spectral efficiency with blocks of 5 MHz of contiguous spectrum, LTE spectrum needs two blocks of 15 or 20 MHz to be most efficient depending. With 5 MHz swaths, the spectrum is approximately 25% as efficient as 20 MHz for LTE services.

Reorganizing spectrum nationally into wider and more efficient swaths requires significant changes to the current system. It is analogous to replacing country roads with a major interstate highway. Ownership of spectrum by many smaller entities impedes the ability to conduct the spectrum swaps necessary to create larger blocks. At the same time, most national regulators are often loath to limit the number of competitors for fear of anti-trust issues.[76]

With a block of spectrum for each of the two transmission directions, the spectrum can be used more efficiently

b. Value of Pairing

In addition to the issue of large contiguous blocks, spectrum often needs to be "paired," i.e. allocated in two pieces, separated enough to avoid interference, but close enough that the propagation characteristics are not materially altered. This allows one block to be used for the "uplink" (from the handset to the base station) and the other for the "downlink" (from the base station to the handset). With a block of spectrum for each of the two transmission directions, the spectrum is used more efficiently. Economist Coleman Bazelon performed a detailed analysis of the value of pairing.[77] Closely spaced spectrum pairs (small duplex gaps) are more desirable because the propagation differences between the uplink and downlink are minimized. However, it is often easier for

---

[76] Regulators in the Netherlands have concluded that greater efficiencies in spectrum sharing can be achieved via a separation of infrastructure providers (who own and build out the spectrum), and service providers (who provision the service).

[77] Coleman Bazelon, The Brattle Group, "The Economic Basis of Spectrum Value: Pairing AWS-3 with the 1755 MHz Band is More Valuable than Pairing it with Frequencies from the 1690 MHz Band," April 11, 2011.

regulators to find larger blocks to allocate if they are further apart (large duplex gaps). As large blocks are generally more valuable than small blocks regulators face a tradeoff.

Some of the latest spread spectrum technologies such as LTE-TDD, however, do not benefit from pairing. So, the enhanced value associated with paired spectrum may decline over time.

3. Hardware Availability Must Be Considered

A significant driver behind efforts at international harmonization is a desire to increase hardware availability

Hardware availability is an important consideration, particularly in price-sensitive applications, including most consumer markets. Operators resist paying high prices for spectrum for which there is little hardware available, or if the hardware market is not price-competitive. They are legitimately concerned that a lack of competitively priced hardware will make it more difficult for them to attract customers or force them to pay greater hardware subsidies to get customers. The challenge of available hardware is more common with small amounts of spectrum. Without adequate spectrum, hardware manufacturers may believe demand will be insufficient even if the spectrum is of high quality. A significant driver behind efforts at international harmonization is a desire to increase hardware availability. Manufacturers are more likely to develop hardware for spectrum if there is an international market as opposed to only a domestic one.

The spectrum crunch is largely an urban phenomenon

On the other hand, operators sometimes seek to strategically limit hardware. In the 700 MHz band, Verizon and ATT have tried to limit handsets so that they do not work on the A Block generally owned by smaller operators. They were seeking to effectively "lock" their hardware to their own spectrum and, at the same time, make it difficult for smaller operators to have cost effective handsets without economies of scale.

4. Geography Is Critical

In general, demand for wireless services is much greater in urban areas. The more populous a given area, the higher the density of use and the more spectrum needed. As a result, spectrum crunch is largely an urban area phenomenon. In fact the FCC indicated in

2010 that the then existing 608 MHz of wireless spectrum in the United States was sufficient for most rural areas.[78] Urban spectrum is, therefore, more valuable than rural spectrum. A license that covers urban areas is likely worth considerably more than one in rural areas. Additionally, licenses for large areas are priced lower than licenses for smaller areas since smaller license territories allow buyers to more narrowly bid for territories that they need most for their businesses. For very large territories, the cost of the required build-out restrains the bidding level. For example, few companies have the resources to finance a national build-out. So a national license should sell for a discount on a price per MHz/PoP basis than the aggregation of smaller regional licenses covering the country.

It would be a mistake to blindly assume that the current scarcity in certain market will continue indefinitely

5. Supply and Demand Considerations

In addition to the current supply-demand dynamics that drive the analysis above, we must also consider future supply and demand. While short-term mismatches are possible, the FCC is moving on several fronts to allocate additional spectrum to the areas most in demand. Although, as described in Section V.A above, the FCC reallocation process can take years, it would be a mistake to blindly assume that the current scarcity in certain markets will continue indefinitely. Regulatory processes, although slow, often have a tendency to overcompensate when they finally begin to move.

In addition to the absolute issues of supply and demand, there are company specific issues to consider. In the U.S. Verizon and ATT are under regulatory and political pressure to reduce or maintain their spectrum at current levels particularly in the lower bands. Current rules, called "spectrum screens", generally limit operators from making further acquisitions once their control more than 33%r of the spectrum in a give geographical area. With these deep-

[78] FCC Staff Technical Paper, "Mobile Broadband: The Benefits of Additional Spectrum," October 2010, available at: http://www.fcc.gov/document/mobile-broadband-benefits-additional-spectrum.

pocketed players out of the bidding in certain spectrum sales, demand may be may be lower, resulting in a reduced price.

6. Auction Rules

Policymakers designing auctions must balance the desire to attract multiple bidders to ensure a competitive auction against the desire that each operator have enough bandwidth to operate efficiently. This is often difficult to achieve.

As a result of the difficulty in balancing these interests, prices in spectrum auctions vary widely. Auctions for identical spectrum in similar countries occurring within weeks of each other can yield price differentials of over an order of magnitude. Details include bidding eligibility that impacts the number of potential bidders, the structure and usage rules of the spectrum and other circumstances related to the specific market.[79]

Auctions for identical spectrum in similar countries can yield price differentials of over an order of magnitude

7. Incumbent vs. New Entrant

In general, spectrum is more valuable to an incumbent than a new entrant. This is because an incumbent can leverage existing infrastructure. Moreover, incumbents view added spectrum as a savings, i.e. an expenditure on additional cell sites avoided. Their incremental revenue per subscriber is also likely to be higher, allowing them to bid more. An established incumbent also has a lower cost of capital. When selling, an incumbent may view the spectrum as part of its going-concern value and require a higher price than it would for an asset. Incumbents may also assign "private value" to the spectrum due to the potential value of using it to block new entrants.

8. License Terms

License terms have a great impact on the value of spectrum. Issues to consider include the term, renewal expectations, transfer rights

---

[79] For a detailed discussion of auction structure and policy issues, see Martyn Roetter, on behalf of the GSMA, "Spectrum Broadband in the Americas" Policy Issues for Growth and Competition," January 2011.

and build-out requirements as well a potential sharing and/or reuse requirements. As a result of the enormous cumulative impact of different national auction processes, license rules, operation regulations and competitive environments, auction results in one country have limited value in projecting auction valuations in another country.

Around the time of the introduction of the iPhone in 2008, mobile broadband spectrum prices soared

H. Valuation Trends

Around the time of the introduction of the iPhone in 2008, mobile broadband spectrum prices soared. The growth of wireless broadband became obvious, but the strategies of operators to deal with it were less obvious. It is now clear that operators have managed the growth via data caps, smaller cell sites, and Wi-Fi offloading, among other strategies. As a result, the upward trajectory in price increases has stalled as shown in Exhibit 6-4 below:

Exhibit 6-4: Consolidated Spectrum Return Indices

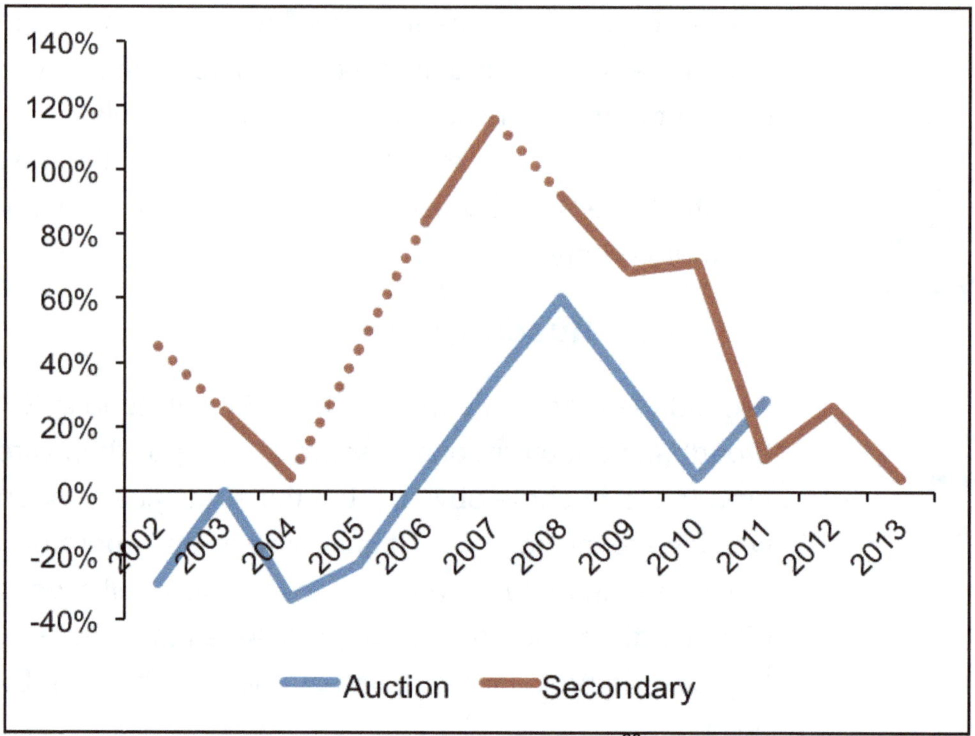

Source: Scott Wallstein, Technology Policy Institute.[80]

The supply side of the spectrum equation is more complex than the demand side

This is consistent with an international study of 200 mobile wireless auctions showing a decline in relative prices for spectrum globally.[81] Nonetheless great uncertainty remains about future spectrum supply and demand.

1. The Supply Debate

The supply side of the spectrum demand equation is more complex than the demand side. Much of the uncertainty is due to the difficulty in predicting the outcome and timing of various government efforts to increase spectrum availability. But the following is known:

---

[80] For explanation of calculations, see Scott Wallsten, Technology Policy Institute, "Is There a Spectrum Crisis? Quantifying Factors Affecting Spectrum License Value," February 26, 2013, available at: http://bit.ly/ZluhUp.

[81] Sims, Martin, "Price of Mobile Spectrum Falls in Real Terms", Policy Tracker, June 20, 2011.

1. The 2010 National Broadband Plan seeks to add 300 MHz of spectrum for mobile broadband by 2015 and 500 MHz by 2020. The largest part of that is the 120 MHz that the FCC seeks to pull back from television broadcasters via the reverse broadcaster incentive auction process.
2. The PCAST Report suggests the government share 1,000 MHz of federal spectrum below 3.7 GHz with commercial users. Nearly 500 MHz of government spectrum, although only about half below 3.7 GHz, is already in the NPRM process for shared use.
3. The FCC is required to auction 65 MHz of additional mobile broadband spectrum by early 2015.
4. There is also a movement to force spectrum license users to allow others to use their spectrum when they are not using it.[82]

The television broadcasting incentive auction is currently attracting a great deal of attention

Cumulatively, these plans, if realized, could massively increase spectrum supply for wireless services.

But, as previously discussed, these processes are behind schedule–some observers say hopelessly. Many observers, although a minority, believe that few government agencies will ultimately agree to share spectrum. Others suggest wireless companies will be reluctant to be "second class citizens" on shared spectrum in large numbers.[83] If so, they will resist investing in the infrastructure needed too widely deploy shared network systems.

---

[82] For a detailed analysis of this proposal see: Michael Calabrese, TPRC 2011, "Use it or Share it: Unlocking the Vast Wasteland of Fallow Spectrum" September 25, 2011, available at SSRN: http://ssrn.com/abstract=1992421.

[83] For a detailed analysis of this position see: J. Zander, L.K. Rasmussen, K. Sung, P. Mahonen, M. Petrova, R. Jantti & J. Kronander, "On the Scalability of Cognitive Radio: Assessing the Commercial Viability of Secondary Spectrum Access," *Wireless Communications, IEEE* , vol.20, no.2, 28, 36, April 2013, available at: http://ieeexplore.ieee.org/stamp/stamp.jsp?tp=&arnumber=6507391&isnumber=6507381

The television broadcasting incentive auction is currently attracting a great deal of attention. Some industry observes believe that it will not happen due to political issues, while others expect it within the next 18 months, releasing 120 MHz of spectrum nationwide into the market. The former group notes that all of the other spectrum reallocation efforts are all years behind schedule. Some among them believe it will fail because the spectrum is too valuable for the broadcasters to sell at prices that the FCC will likely offer. One leading expert has told Summit Ridge Group that the broadcast incentive auctions is in doubt because so much of the government spectrum is being opened for sharing per the PCAST recommendations. Nearly 500 MHz of shared spectrum allocation is currently in the NPRM process – 3.5GHz (100-150 MHz), 4.9GHz (50 MHz), 5Ghz (195 MHz) and potentially in the 1750-1850 MHz (100 MHz) band. This camp argues that the incentive auction may be called off because the broadcasters' spectrum simply will not be needed. (Spectrum supply increased is discussed in further detail in Section V above).

The Aereo, Inc. situation creates an additional wrinkle for potential spectrum supply

The Aereo, Inc. Wrinkle

The Aereo, Inc. situation creates an additional wrinkle for potential spectrum supply. Some broadcasters (including Fox) are now publically indicating that if the Second Circuit decision affirming Aereo's right to sell its over-the-air broadcasts through the internet, they may stop broadcasting over the air and only sell to satellite and cable systems. If this happens, television broadcasters will have little incentive to keep their spectrum and would eagerly participate in the auction process. This could increase spectrum available for reallocation to mobile broadband. Ironically, in this case it would not be the weakest broadcasters who would participate, but rather the strongest. The strongest broadcasters, especially those with network affiliates, have leverage with the cable companies. The smaller ones are more likely to be dependent on "must carry" rules that are tied to broadcasting over the air.

Of course the Aereo's legal situation is not completely clear. A lower level (district court) in the Ninth Circuit (covering nine western states plus Guam and the Northern Mariana Islands) ordered a similar service to shut down. The judge indicated that the broadcasters were likely to win their claim of copyright infringement. Should the case be appealed and the Ninth Circuit rules opposite the Second Circuit, it would set the stage for the Supreme Court to decide, possibly years from now. The broadcasters may also seek regulatory relief from Congress to simply outlaw Aereo's business model and circumvent the current court battle. Although broadcasters have much influence in Washington, the government is encouraging them to cede their spectrum. Passing a law strengthening their hold on spectrum would seem inconsistent.

There is little consensus about the rate of future demand growth for spectrum

2. The Demand Debate

Like the argument with respect to supply, there is little consensus about the rate of future demand growth. The issue is more fundamental – predicting consumer usage patterns and improvements in spectrum efficiency. Some experts point to high historical demand growth and the increased rollout of smart phones, which use a multiple of the amount of spectrum of other phones. The introduction of 4G devices will further increase internet usage, especially with media downloading. Cisco, for example, projects a 10-fold growth in mobile demand from 2012 and 2017. Currently, the FCC has allocated approximately 608 MHz to wireless broadband. The "spectrum crunch" side argues that such demand growth will outpace any likely increase in supply.

Others say the demand growth is exaggerated.[84] They point to data caps that most operators have put on their new plans. Some evidence shows that the growth in data traffic has slowed. Newer 3G and 4G protocols are far more spectrally efficient than earlier

---

[84] Tim Farrar, "Is Cisco Stacking the Deck with its Mobile Data Numbers?" Gigaom, accessed May 15, 2013, http://gigaom.com/2013/02/09/is-cisco-stacking-the-deck-with-its-mobile-data-numbers.

versions. Additional rapid Wi-Fi off-loading – estimated at close to 33% of wireless traffic in 2011 and growing to as much as 60% today — is absorbing much of the demand growth. Skeptics of the "spectrum crunch" point to a study cited in the PCAST report indicating that less than 20% of the capacity in prime spectrum bands under 3.7 GHz is utilized even in the most congested areas.[85] The PCAST Report argues that more efficient spectrum use could increase effective capacity by a factor of 1,000.[86]

The lack of consensus attracts investors because it creates an opportunity to make money for those who can get it right

3. Opportunities

It is rare for an industry as mature as the U.S. wireless industry to have such little consensus about medium-term supply and demand growth. But a lack of consensus attracts investors because it creates an opportunity to make money for those who can get it right. Currently investors are acquiring smaller broadcasters in major markets with the hope of selling the acquired broadcaster's spectrum back to the FCC at an even higher price. Broadcasters expecting the auction to be less attractive are eagerly selling to them. Wireless companies are making major spectrum transactions with little industry certainty about future supply that could significantly impact the values of the spectrum that they are trading.

---

[85] See PCAST Report, 99.

[86] See PCAST Report, iv.

VII. Global Perspectives

The United States is not alone in its need for additional spectrum for mobile data applications. As of the end of 2012, the global demand for mobile data was twelve times greater than in 2000 and Cisco projects a thirteen-fold increase between 2012 and 2017.[87] This growth is ubiquitous around the globe regardless of geographic, political and socio-economic factors. While almost every country has its own policies to address spectrum issues, most have adopted or are considering the approaches that the United States has taken to handle the challenge.

Governments and communications service providers around the world recognize that a global standard will lead to economies of scale of equipment that will reduce costs and increase spectral capability. By harmonizing frequency uses advanced wireless technology (most likely to be LTE) will be able to function in different markets without adaptions, providing significant economies of scale. If global networks use similar technology and standards, there will be more competition among equipment developers to offer quality products at cheap prices. Unfortunately, harmonization has lagged spectrum allocation for mobile broadband, making the former even more difficult to achieve.

Although only a minority of countries have harmonized spectrum, in major markets the most valuable spectrum has already been or is currently being harmonized

Although only a minority of countries have harmonized spectrum, in major markets the most valuable spectrum has already been or is currently being harmonized, primarily in the 700 MHz "digital dividend" band in the 698 to 806 MHz range. For ITU Region 2 (the Americas) and for nine countries in ITU Region 3, including China,

[87] Cisco, "Cisco Visual Networking Index: Global Mobile Data Traffic Forecast Update 2012-2017," 2013, available at: http://www.cisco.com/en/US/solutions/collateral/ns341/ns525/ns537/ns705/ns827/white_paper_c11-520862.html. Hereinafter referred to in text as "Cisco Visual Networking Index."

India, Japan and South Korea, the World Radio Congress (WRC-07) targeted this band for reallocation from television broadcasters after the industry shifted to digital. In the 700 MHz band there are three different band plans options proposed by the ITU.

A. General Trends

The United States and Germany are the global leaders in spectrum allocation to mobile broadband

Governments globally have increased their effort to regulate spectrum with the recognition of the importance of mobile communications for economic development. Each country's ability to influence hardware providers has led to varied strategies.

Developed countries have been more successful at allocating additional spectrum for mobile data, the most pressing area (see Exhibit 7-1). The United States and Germany are the global leaders in spectrum allocation to mobile broadband.

Exhibit 7-1: MHz of Spectrum Allocated to Mobile Operators by Country Globally

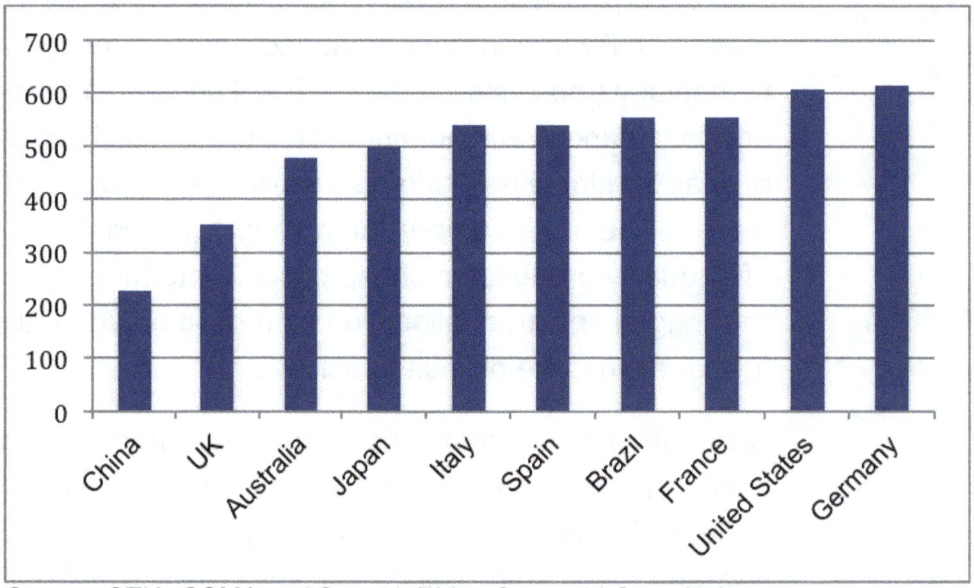

Source: CTIA, GSMA and Summit Ridge Group, LLC analysis.

Anticipated spectrum license allocations for mobile broadband will narrow the gaps between countries as shown in Exhibits 7-2 below.

Exhibit 7-2: Licensed Spectrum Available for Mobile Broadband (MHz)

Country	Current	Pipeline	Current+ Pipeline
U.S.*	608	55+	663+
Australia	478	230	708
Brazil	554	0	554
China	227	360	587
France	555	50	605
Italy	540	20	560
Japan	500	10	510
Spain	540	60	600
U.K.	353	265	618
*US pipeline does not include the significant amount of spectrum that may be made available from incentive broadcast auctions and federal repurposing.			

Source: FCC[88] and Summit Ridge Group, LLL analysis.

Historically, large developed countries such as the United States have been less compliant with international norms

2. Greater Influence of Standards on the Way

Historically, large developed countries such as the United States have been less compliant with international norms. Regulators and service providers in these countries have felt less pressure to adapt to international standards, knowing that manufacturers must accommodate them. Smaller countries like Canada more readily adopt standards set by larger neighbors to manage interference and benefit from economies of scale that these standards provide in equipment manufacturing. Some countries have been pulled in two directions. Mexico is under pressure to adopt the standards of its major trading partner, the United States, against countervailing pressure to adopt those of fellow Spanish-speaking Latin American countries, which also account for significant trade and political influence. Similar dynamics play out around the world.

Recently, the influence of global standards has become stronger.

[88] FCC White Paper: The Mobile Broadband Spectrum Challenge: International Comparisons. Wireless Telecommunication Bureau, Office of Engineering & Technology. February 26, 2013.

Even large countries find it more difficult to act independently. In 2012, foreign operators increased their stakes in the United States market. Japan's Softbank is seeking to acquire Sprint[89], while Deutsche Telecom's increased its stake its exposure to the U.S. market by its T-Mobile subsidiary's October 2012 announced acquisition MetroPCS. Summit Ridge Group believes that these deals were intended, in part, to increase leverage and economies of scale with equipment manufacturers. They also tie the largest market in ITU Region 2 (the Americas) to ITU Regions 1 and 3. As a result, we expect greater international coordination across the board in the near future.

B. ITU Region 1: Europe, Middle East and Africa

1. Europe

Europe has taken the lead, from a regulatory planning perspective, in clearing spectrum for advanced LTE services. In November 2012, the European Commission ordered 120 MHz of spectrum in the 2 GHZ band, previously authorized for 3G services, to be cleared for 4G services by June 30, 2014. Once its current clearing processes are complete, Europe's Radio Spectrum Policy Program (RSPP)[90] should make over 1200 GHz of spectrum available for 4G services by 2020—roughly double the amount available in the U.S. today. The U.S. also plans to reallocate an additional 500 MHz of spectrum for mobile broadband services, but those efforts are behind schedule. Exhibit 7-3 below illustrates the state of mobile broadband spectrum in Europe as of October 2012. At this point, Europe, with the exception of Germany, has less mobile spectrum available than the United States.

---

[89] New York Times, "Confidence From Chief of Softbank in Sprint Bid", June 21, 2013. Available at: http://www.nytimes.com/2013/06/22/business/global/confidence-from-chief-of-softbank-in-sprint-bid.html?pagewanted=all

[90] European Commission, "RSPP: The Roadmap for Wireless Europe," available at: https://ec.europa.eu/digital-agenda/en/rspp-roadmap-wireless-europe.

Exhibit 7-3: Licensed Mobile Broadband Spectrum in Select EU Countries

Band	Germany		Spain		France		Italy		U.K.	
	Curr.	Pipeline	Curr.	Pipeline	Curr.	Pipeline	Curr.	Pipeline	Curr.	Pipeline
< 700 MHz	-	-	-	-	-	-	-	-	-	-
700 MHz	-	-	-	-	-	-	-	-	-	-
800 MHz	60	-	-	60	60	-	60	-	-	60
900 MHz	70	-	70	-	70	-	70	-	70	-
1500 MHz	-	-	-	-	-	-	-	-	-	-
1700/1800 MHz	140	-	150	-	150	-	120	20	143	-
1900 MHz	35	-	20	-	15	-	20	-	20	15
2.1 GHz	120	-	120	-	120	-	120	-	120	-
2.3 GHz	-		-	-	-	-	-	-	-	-
2.6 GHz	190	-	180	-	140	50	150	-	-	190
TOTAL	**615**	**0**[91]	**540**	**60**	**555**	**50**	**540**	**20**	**353**	**265**

Source: FCC and Summit Ridge Group, LLC analysis.

As in the U.S., Europe is also exploring the use of unlicensed spectrum for mobile broadband. Exhibit 7-4 provides a comparison of European and U.S. unlicensed broadband spectrum.

[91] At a recent conference hosted by Hogan Lovells in May 2013, a respected German telecom attorney disputed the accuracy of this FCC figure suggesting Germany has no additional mobile spectrum in its pipeline.

European countries have different agendas for allocating spectrum

Exhibit 7-4: Amount of Spectrum Below 6 GHz Available for Unlicensed Broadband Use

	UNITED STATES		EUROPE	
BAND	Current	Pipeline	Pipeline	Pipeline (Unknow
TV White Spaces	0-150	*	-	
863-870 MHz	-	-	7	
902-928 MHz	26	-	-	
1880-1930 MHz	10	-	20	
2400-2483.6 MHz	83.5	-	83.5	
3550-3700 MHz	50	100	-	
5150-5350 & 5470-5825 MHz	555	-	555	
5350-0 & 5850-5925 MHz	-	195	-	
*Includes "licensed-light" spectrum.				

Source: FCC[92] *and Summit Ridge Group, LLL analysis.*

Europe has a unique challenge in that EU leaders seek to create a unified wireless telecom market with standardized prices at the wholesale level. However, the EU countries allocate spectrum at a national level. Different countries have different agendas for allocating spectrum. Some use it for revenue, while others want to make it available to as many people as cheaply as possible. Some governments focus on innovation by encouraging competition; others prefer fewer competitors to ensure more efficient spectrum use. But, these varying strategies of spectrum allocation are inconsistent with the uniform wholesale pricing the EU leaders seek.[93] Furthermore, as newer LTE technologies require larger spectrum blocks to be efficient, the trade-off between competition and efficiencies becomes even more complex. Despite these challenges, the European Conference of Postal and Telecommunication Administrations [CEPT] has ensured strong technical coordination within the EU. Moreover, the EU has the

---

[92] FCC White Paper, Wireless Telecommunication Bureau, Office of Engineering & Technology, "The Mobile Broadband Spectrum Challenge: International Comparisons," February 26, 2013.

[93] For a more detailed discussion of this topic, see http://coleago.wordpress.com/2013/03/14/towards-a-single-eu-telecoms-market.

power to force its members to comply with EU policies to which their countries are signatories, which include most communications regulation.

Another challenge European operators face is different regulatory regimes in each country. Additionally, there is greater uncertainly over long-term spectrum rights than in the U.S.. These challenges makes cross-border consolidation and achieving economies of scale difficult, Consequently, the two largest operators in the U.S. (AT&T and Verizon) are each larger than the three largest European operators combined.[94]

With respect to investment, Europe has experienced lower levels of capital expenditure over the past decade compared to North America. A recent Goldman Sachs research report[95] indicates that European capital expenditures have declined slightly since 1997. During the same period capital expenditure has increased by approximately 75% in the U.S. The majority of the older mobile networks in Europe are now running out of spectral capacity, resulting in decreasing revenues to the wireless companies, similar to the phenomenon with fixed networks. Meanwhile the newer networks based on more recently added frequencies lack sufficient equipment to facilitate growth that meets the demand for spectrum.

Aggregate wireless capital expenditures in Europe are not keeping pace with those in the United States

Recently, however, capital expenditure has increased in Germany, offering the potential for spectral efficiency gains there. Deutsche Telekom AG will increase capital expenditure by approximately €10 billion over the next three years, with the expectation that the number of customers reached by LTE and Fiber to the Curb (FTTC) will increase to 85% and 65% of total customers respectively by 2016.[96] Despite signs of progress in Germany, aggregate wireless

---

[94] GSMA, "Mobile Wireless Performance in the EU and the US" (May 2013) at p.26. Available at: http://www.gsmamobilewirelessperformance.com/ . Hereinafter referred to in text as "GSMA Mobile Wireless Performance Report."
[95] GSMA Mobile Wireless Performance Report at p.17.
[96] Alan Weissberger, The Viodi View, "Global Telco Capex May Surprise on the Upside in 2013!," December 25, 2012, see http://viodi.com/2012/12/25/global-telco-capex-may-surprise-on-the-upside-in-2013.

capital expenditures in European are not keeping pace with those in the U.S. A Vodafone spokesman recently attributed the over-fragmentation of the European market to the lower returns on investment in Europe, explaining the lower capital expenditures. These results are supported by a McKinsey & Company report[97] which also blames regulations requiring operators to sell capacity at low wholesale rates and to unbundle services not found in the U.S.

2. Middle East

Mobile wireless in the Middle East is growing rapidly, but spectrum dynamics vary widely there due to the wide variety of political regimes and levels of economic development. For this reason, generalizations are difficult. However, the comparatively young population of the region is a key factor driving increased demand for mobile broadband service. Several multinational wireless operators have emerged that may influence greater uniformity in the region's telecom industry over time.

One major difference between the Middle East and the OECD ITU regions it that the former is characterized by high rates of pre-paid users

One major difference between the Middle East and the OECD ITU regions is that the former is characterized by high rates of pre-paid users. Users of pre-paid SIM cards are not tied to contracts and can easily change providers. As a result of low switching costs, the Middle East market is extremely price-competitive. This may explain why mobile operators are leaders at mobile advertising,[98] a technique to enhance revenue without the need for increasing subscriber rates.

3. Africa

Africa is a highly underserviced continent from a telecommunications perspective. In many areas landline telephony is non-existent or poorly developed. Investors have avoided the

---

[97] McKinsey & Company, Telecommunications, Media and Technology Practice, "A 'New Deal': Driving Investment in Europe's Telecoms Infrastructure," November 2012.

[98] Alawaba Business, "Middle East Spending on Mobile Ads Goes Through the Roof, December 9, 2012, accessed May 15, 2013, available at: http://www.albawaba.com/business/mideast-mobile-ads-455486.

region due to its poverty and concerns over corruption and its overall business climate. However, over the past 10-15 years, many African countries have increased the transparency of their regulatory regimes and become more open to foreign investment. The African telecom market has been proactive in improving its networks. As a result, capital expenditure for telecommunications hardware and software has been steadily increasing. With low levels of existing terrestrial infrastructure, service providers often bypass terrestrial telecom systems to transmit signal across its final segment to the customer (last mile access) with wireless systems.

C. ITU Region 2: The Americas

1. United States Faces Specific Spectrum Challenges

In the United States, the spectrum crunch is a result of events that occurred in the early history of the U.S. telecommunications industry. The United States was early in providing spectrum licenses so more licenses have been issued. Unlike many European countries, the United States has a large commercial broadcasting industry, which has made huge private investments to develop the spectrum covered by its licenses. This investment makes it politically difficult for the government to reclaim and reallocate spectrum for higher uses. The U.S. government, particularly the military, also controls large segments of usable spectrum. As both the commercial broadcasters and government users are hard to move, the U.S. the domestic spectrum challenge is particularly difficult to tackle.

The U.S., due to the responsiveness of the FCC, has led other countries in coping with the spectrum crunch

Despite the challenges it faces, the United States, due to the responsiveness of the FCC, has led other countries in coping with the spectrum crunch. Thus U.S. spectrum policy now influences global spectrum policy such as the allocation for mobile use as shown in Exhibit 5-1 in Section V. The United States also has been leading the movement towards shared usage and the regulation of white space devices (WSD). Currently, approximately 70% of LTE subscribers worldwide are in the United States.[99] Of course, this lead is likely to rapidly shrink as other countries roll-out their LTE systems.

The FCC appears to be proactive in the dialogue with other governments and international organizations to encourage the harmonization of spectrum allocations to wireless broadband. The next World Radio Communication Conference (WRC), hosted by the United Nations in 2015, will provide a forum for this.[100]

COFETAL, the Mexican telecom regulatory agency, announced that Mexico would adopt the APT band plan for the 700 MHz band, followed, by the rest of Latin America. This effectively splits the ITU Region 2 into two different standards

In addition to regulatory efforts, wireless sector investment has been strong. AT&T has announced raising its capital expenditure by a multi-billion dollar sum. It will invest about $14 billion over the next few years in its "Project Velocity." A majority of the money for this initiative will be allocated to improving wireless services expansions including LTE networks. As previously mentioned, international operators like Deutsche Telecom and Softbank announced significant investments in the U.S. wireless market in 2012. This suggests that industry players see strong growth potential in the US market.

2. Canada and Mexico

As mentioned above in Section VII.A., Canada and Mexico are under pressure to accommodate the U.S. band plans and other regulations, ostensibly for the convenience of users and to further the drive for economies of scale with hardware manufacturers who service the U.S. market. Mexico, however must also consider the influence of its Latin American neighbors.

In late 2012, COFETEL, the Mexican telecom regulatory agency, announced that Mexico would adopt the APT band plan for the 700 MHz band, followed by the rest of Latin America and Asia, but at odds with the U.S. band plan. This effectively splits ITU Region 2

[99] The U.S. lead is not necessarily considered a good thing. One European regulator has said, "We [in Europe] watch the FCC race ahead [with a certain regulatory issue], then we examine their mistakes and do it right ourselves a few years later."

[100] There is some debate about shifting this meeting to 2016.

into two different standards for the 700 MHz band. Given the importance of the 700 MHz band, this signals a decline in U.S. influence over its regional partner.

3. Latin America

Spectral efficiency has been a more pressing issue in Latin America than in the United States and Europe. There, relatively little spectrum has been allocated to mobile broadband services. Brazil is the only Latin American country with more than 400 MHz allocated to mobile services (compared to 608 MHz in the United States). Most Latin American countries have less than 300 MHz allocated, causing concerns about the lack of spectrum and spectral efficiency. Given the less developed state of fixed access networks there, the wireless networks are even more important. If more spectrum cannot be opened up for additional mobile broadband networks, then Latin American countries will find it very difficult to meet their targets for economic and social development, which include providing the maximum number of potential users with accessible wireless networks of the highest capacity and quality.

Most Latin American countries have less than 300 MHz allocated [to mobile broadband], causing concerns about the lack of spectrum and spectral efficienty

Underutilized frequency bands exist in Latin America as they do in the United States. However, plans for most of these bands are not well developed there so there will be significant delays before these frequencies can effectively support broadband services. Initial steps have been taken in Chile and Mexico for underutilized spectrum bands to be used to provide broadband services, including the AWS 2.5 GHz, 700 MHz, 850 MHz and 1900 MHz bands.

In addition to opening up underutilized frequencies, industry stakeholders must consider the structure of the spectrum potentially reallocated for mobile wireless services. One specific issue is the width of the frequency blocks for mobile wireless services. Making broader blocks available for certain uses, such as LTE networks, will increase network efficiency. A 15 MHz block of spectrum is several time more efficient for LTE use than a 5 MHz block. Using larger blocks therefore effectively decrease the costs of using the network.

As part of the global harmonization effort to coordinate frequency usage, the United States has, over the years, entered dozens of bilateral and regional agreements with Canada and Mexico that set the terms for sharing spectrum in the United States and near the border regions.

D. ITU Region 3: India and Asia

The Asian-Pacific market is forecasted to account for a third of global mobile services revenue by 2016.[101] Companies and governments there are particularly concerned about developing spectrum-efficient ways of operating mobile services.

Operators in the region have been making more consistent capital expenditures to improve spectral usage than companies in Europe, North America and Latin America. Interestingly, major Asian countries are now beginning to allocate the 700 MHz band to mobile wireless, a move that the U.S. recently completed.

China... has elected to primarily use time division duplexing (TDD) for its LTE standards as opposed to the frequency division duplexing (FDD)

1. India

India spectrum regulation has a checkered history. It granted 2G licenses in the 900 MHz spectrum in 2008, but the courts cancelled them in early 2012 over concerns of corruption. It then allowed unsuccessful bidders to offer service in that band, only to reverse itself and order a spectrum auction for 2013. This history has made international investors reluctant to enter the Indian market.

Since the tainted 2008 auction, India has held two successful spectrum auctions including a 2010 auction for 3G and 4G services and 2012 auction for 2G services in the 800 and 1800 MHz bands.

2. China

China, one of the fastest growing economies, has elected to use

---

101 Alan Weissberger, The Viodi View, "Global Telco Capex May Surprise on the Upside in 2013!," December 25, 2012, see http://viodi.com/2012/12/25/global-telco-capex-may-surprise-on-the-upside-in-2013.

time division duplexing (TDD) for its primary LTE standard as opposed to the frequency division duplexing (FDD) which the U.S. predominately uses. This may provide greater flexibility as it will not need separate frequency bands for uplink and downlink. This flexibility is likely to be important as China has only allocated 227 MHz of spectrum to mobile broadband. China Mobile, however, indicates it will also operate an FDD network in parallel to its TDD network.[102]

3. Japan

As shown in Exhibit 7.5 below, Japan has allocated 500 MHz of spectrum to mobile broadband use with 10 additional MHz in the pipeline for use in the next three years.[103] Although less aggregate spectrum is available than in other developed countries, the range of frequencies allocated to mobile broadband is quite broad. Japan is also one of the few countries, in addition to the U.S., to have already allocated spectrum in the 700 MHz band for mobile broadband. It is also one of the few countries to have allocated spectrum in the 1500 MHz band for mobile broadband.

---

[102] Benny Har-Even, Telecoms, "China Mobile to Operate TDD and FDD LTE in Tandem CEO Reveals,"January 22, 2013, available at http://www.telecoms.com/75811/china-mobile-to-operate-tdd-and-fdd-lte-in-tandem-ceo-reveals.

[103] FCC International Spectrum White Paper, 7.

Exhibit 7-5: US v. Japan – Licensed Mobile Broadband Spectrum as of October 2012

Band	United States		Japan	
	Current	Pipeline	Current	Pipeline
< 700 MHz	-	?	-	-
700 MHz	70	-	60	-
800 MHz	64	-	60	-
900 MHz	-	-	30	-
1500 MHz	-	-	70	-
1700/1800 MHz	-	15	70	10
1900 MHz	130	10	30	-
2.1 GHz	130	30	120	-
2.3 GHz	20		-	-
2.6 GHz	194	-	60	-
TOTAL	**608**	**0**	**500**	**60**

Source: FCC and Summit Ridge Group, LLC analysis.

If Japan-based Softbank completes its acquisition of Sprint [104] it should add pressure for harmonization of U.S. and Asian spectrum policies, particularly for hardware and handset standards. Harmonization would likely increase economies of scale and purchasing power for wireless operators' over their suppliers.

4. South Korea

South Korea is one of the most advanced countries for communications in the world. According to the Korean regulator KCC, mobile data traffic on 2G, 3G, and 4G networks increased approximately 80% between January and November of 2012.[105] Korea Telecom, LG U+ and SK Telecom collectively provide nearly 100% of the population with mobile services.

---

[104] DISH Networks' counterbid for Sprint creates uncertainty over Softbank's acquisition. See Kevin Fritchard, "Sprint's Tough Choice: Dish Might Be a More Attractive Suitor than Softbank," April 15, 2013, available at http://gigaom.com/2013/04/15/sprints-tough-choice-dish-might-be-a-more-attractive-suitor-than-softbank.

[105] Cisco Visual Networking Index

VIII. Economic and Regulatory Conclusions

Making more spectrum available and sharing bands to increase efficiency would ease the spectrum supply challenge and improve the quality of wireless services – both in the U.S. and internationally. This would also have a positive macroeconomic effect. Increasing the amount of spectrum available to commercial entities will stimulate growth by creating jobs and making business more efficient, and also improve the quality of life and competitiveness of each country's economy.

Summit Ridge Group believes that the risk of the "spectrum crunch" turning into a full-blown crisis is becoming remote despite various uncertainties

In the U.S., the FCC has been working aggressively to reallocate spectrum for uses that are in the highest demand. With multiple processes in place, it has granted a number of license waivers and made other regulatory accommodations to facilitate the reallocation of spectrum to the areas of the most vital need. Additionally, nearly 500 MHz of government spectrum is in the Notice of Proposed Rulemaking ("NPRM") stage for shared use applications. Wireless operators have also learned to do more with less spectrum. As a result, Summit Ridge Group believes that the risk of the "spectrum crunch" in the U.S. turning into a full-blown crisis is becoming remote. We acknowledge however, that this is not the consensus opinion among industry experts given the uncertainties over the timing of the reallocation process and demand projections. Few countries are close to the U.S. in terms of allocation additional spectrum to mobile broadband, but most developed economies are making significant progress.

PART THREE: SPECTRUM BAND DESCRIPTIONS AND ANALYSIS

IX. Mobile Wireless Service Band Descriptions

The lower frequency allocations (under 1 GHz) are generally more valuable due to their longer distance propagation and improved building penetration capacity

Mobile Wireless services operate in multiple portions of the radio spectrum between 698 MHz and 2.690 GHz. The lower frequency allocations (under 1 GHz) are generally more valuable due to their longer distance propagation and improved building penetration capacity. They are the most efficient way for wireless service providers to cover large rural areas. In the United States, mobile spectrum licenses are awarded for a period of 10 years with the expectation of renewal. Additionally, white space service providers (described in Section IV.C) operate without licenses in unused broadcast spectrum. Spectrum sharing technology is increasingly feasible but, aside from Wi-Fi, currently constitutes a very small portion of mobile broadband traffic.

A. Upper and Lower 700 MHz bands

The 700 MHz spectrum, running from 698 MHz to 806 MHz, is a relatively low frequency with a correspondingly high wavelength capable of overcoming obstacles in the environment better than higher frequencies. It is currently the most sought after frequency for mobile wireless applications. The long propagation range of 700 MHz spectrum (as well as 800 MHz spectrum described below) is ideal for providing wireless service in rural areas.

The 700 MHz band is currently the most sought after frequency for mobile wireless applications

Due to the long propagation range of the frequency, service providers need fewer base stations to build a network for this frequency range. The cost of building the network is therefore less expensive than for those with frequencies with shorter propagation range. Another attractive feature of the 700 MHz spectrum band is that antennas for this frequency range are small enough to fit into handheld devices like cell phones. The lower cost of building-out the spectrum is off-set by the lower system capacity offered by a network with fewer base stations. As a result, the highest demand for 700 MHz spectrum is in areas with lower traffic volume and/or less bandwidth intensive applications including M2M services.

1. US 700 MHz Allocation

Many other countries around the world are in the process of a reallocation of the 700 MHz band from television to mobile services

Until 2008, the lower portion of the 700 MHz spectrum, from 698 MHz to 746 MHz, was used for television channels 52 to 59. Exhibit 9-1 below illustrates the lower portion of the 700 MHz band plan. The upper portion, from 746 to 806 MHz, was previously used for channels 60 to 69 (as illustrated in Exhibit 9-2). In the U.S., the upper and lower 700 MHz band have been available for mobile wireless service since 2008 when the FCC digitalized and repacked channels 52 to 60 into channels 2 to 51 in the 600 MHz band and auctioned the spectrum to mobile wireless operators. Many other countries around the world are in the process of a similar reallocation of this band from television to mobile services. The lower 700 MHz band is often referred to as the "digital dividend" spectrum. Its reallocation from television to mobile wireless was accompanied by a shift in television broadcasting to digital transmission, a technology requiring less spectrum. Parts of the 700 MHz band, previously allocated for television channels 62-64 and 67-69 have also been allocated for a nationwide public safety network dubbed "FirstNet."[106] The U.S. Congress has authorized $7 billion to construct the FirstNet system.

[106] For information about FirstNet, see FirstNet page of the National Telecommunications & Information Administration Website. Available at http://www.ntia.doc.gov/category/firstnet.

Exhibit 9-1: U.S. Lower 700 MHz Band Plan

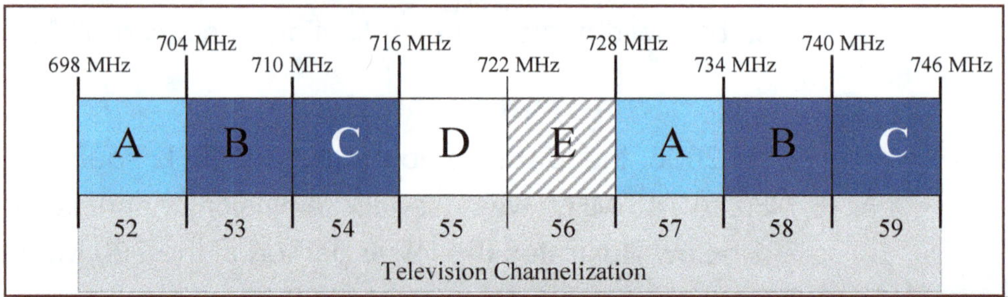

Block	Frequencies (MHz)	Bandwidth	Paring	Geographical Area Type	No. of Licenses
A	690-704, 728-734	12 MHz	2 x 6 MHz	700 MHz EAG	6
B	704-710, 734-740	12 MHz	2 x 6 MHz	700 MHz EAG	6
C	710-716, 740-746	12 MHz	2 x 6 MHz	MSA/RSA	734
D	716-722	6 MHz	Unpaired	700 MHz EAG	6
E	722-728	6 MHz	Unpaired	700 MHz EAG	6

Source: FCC and Summit Ridge Group, LLC analysis.

Exhibit 9-2: U.S. Upper 700 MHz Plan

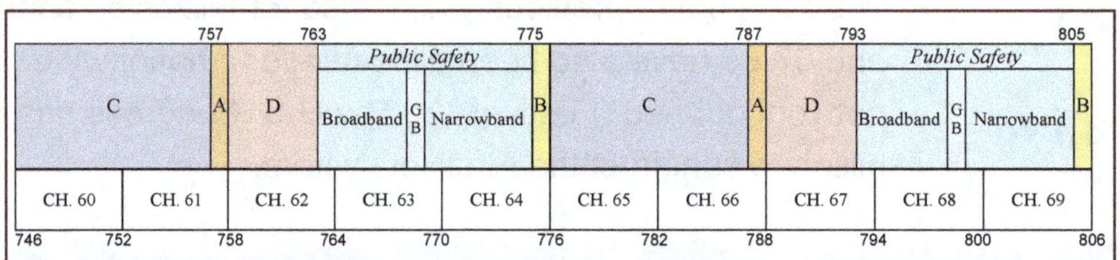

Block	Frequencies (MHz)	Bandwidth	Paring	Geographical Area Type	No. of Licenses
C	746-757, 776-787	22 MHz	2 x 11 MHz	REAG	12
A	757-758, 787-788	2 MHZ	2 x 1 MHz	MEA	52
D	758-763, 788-793	10 MHz	2 x 5 MHz	Nationwide	1*
B	755-776, 805-806	2 MHz	2 x 1 MHz	MEA	52
*Subject to sharing with adjacent public safety network band					

Source: FCC and Summit Ridge Group, LLC analysis.

2. Criticisms of the U.S. Band Plan

The FCC's proposed band plan for 700 MHz faces several criticisms

Although the United States is at the forefront of developing the 700 MHz band for mobile use, the FCC's proposed band plan for 700 MHz faces several criticisms. First, the 700 MHz band is highly fragmented so operators often cannot acquire the large 15 to 20 MHz blocks needed for the efficient use of advanced technologies such as LTE. The A Block in the lower portion of the band is separated from Channel 51 of the television band by only a 1 MHz gap, causing significant interference that severely limits its use. In the 2008 auction, the A Block sold, on average for approximately 60% less than the B Block, which experiences no such interference. Much of the A Block is owned by second-tier service providers. The largest service providers have resisted ordering handsets that operate in the A Block. The lack of handsets that function in the A Block makes it harder for the second tier players to operate and helps to reinforce the dominant position of the larger players.

3. APT 700 MHZ band Plan

The leading alternative to the U.S. band plan for the 700 MHz spectrum is the plan proposed by the Asian Pacific Telecommunity (APT). The APT band plan calls for either 90 MHz of paired frequency-division duplex (FDD) spectrum (2 x 45 MHz) with a 10 MHz duplex gap in the center, or 108 MHz of time-division duplex (TDD) spectrum. FDD is a system of dividing the send and receive signals into different frequencies. TDD, by contrast, uses the same frequencies for sending and receiving, but alternates using them at different times to avoid interference. The APT plan is illustrated in Exhibit 9-3 below.

Exhibit 9-3: APT 700 MHz Band Plan

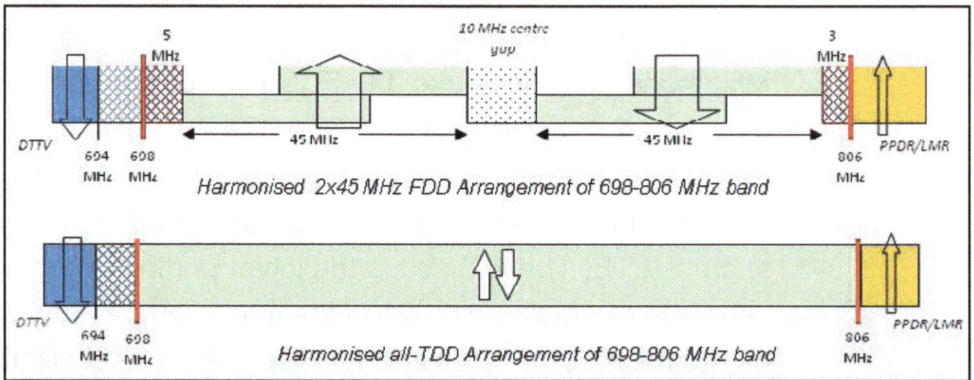

Type	Frequency Range	Amount	Use
APT - FDD	698 to 703 MHz	5 MHz	Guard Band
APT – FDD	703 to 748 MHz	45 MHz	Uplink
APT – FDD	748 to 758 MHz	10 MHz	Centre Gap
APT – FDD	758 to 803 MHz	45 MHz	Downlink
APT – TDD	803 to 806 MHz	3 MHz	Guard Band
APT – TDD	694 to 698 MHz	4 MHz	Guard Band
APT – TDD	698 to 806 MHz	108 MHz	Two-way TDD

Source: 4G Americas and Summit Ridge Group, LLC analysis.

The APT plan has become popular in Asia and Latin America as it avoids the fragmentation inherent in the U.S. plan. However, using TDD and FDD on the same frequency is problematic since TDD uses the same frequency on the uplink and downlink, which interferes with FDD. Because few government regulators seem prepared to choose one or the other, Summit Ridge Group believes that both will be used in adjacent areas, causing additional fragmentation.

B. Advanced Mobile Phone Systems (AMPS): 800 MHz

The original wireless system, AMPS, is a spectrum band located at 824 MHz to 849 MHz and 869 to 894 MHz

The original wireless system, AMPS, is a spectrum band located at 824 MHz to 849 MHz (uplink to base station) and 869 to 894 MHz (downlink to handset). It was originally used for analog (1G) wireless phone services. The FCC allocated the AMPS band into two blocks (the "A block" and the "B block") in each of the 734 Cellular Market Areas (CMAs) in the United States. Each block originally contained 333 channels, expanded in the late 1980s to

416 channels after spectrum was reallocated from UHF TV channels 70-83.

The FCC gave the B block to the local Bell operating companies while it allocated the A Block, at no cost, in a comparative hearing to parties that it judged would put it to superior use. In most CMAs, the B Block licenses went to AT&T. The major license holders of this spectrum are still AT&T and Verizon (successor of NYNex). The FCC sold licenses for several remaining CMAs in Auction No. 45 in 2002 and Auction No. 77 in 2008. A summary of the frequency allocations is in Exhibit 9-4 below:

Exhibit 9-4: Former 800 MHz Cellular Channel License Blocks

A Block		B Block	
Spectrum	**Amount**	**Spectrum**	**Amount**
824.0 to 835.0	11.0 MHz	835.0 to 845.0	10.0 MHz
845.0 to 846.5	1.5 MHz	846.5 to 849.0	2.5 MHz
869.0 to 880.0	11.0 MHz	880.0 to 890.0	10.0 MHz
890.0 to 891.5	1.5 MHz	891.5 to 894.0	2.5 MHz
25 MHz total		**25 MHz total**	

Source: FCC and Summit Ridge Group, LLC analysis.

As of 2008, U.S. wireless services providers were allowed to terminate their analog services and switch entirely to more spectrum efficient digital services

As of 2008, U.S. wireless services providers were allowed to terminate their analog services and switch entirely to more spectrum efficient digital services. All major providers now use this spectrum to provide services using digital technologies.

The FCC paired the 804 to 824 MHz section with the 851 to 869 MHz section of the band and allocated it for various public safety radio systems used by police and emergency responders. See Exhibit 9-5 below for a graphic description of the 800 MHz band:

Exhibit 9-5: 800 MHz Plan – Post reconfiguration

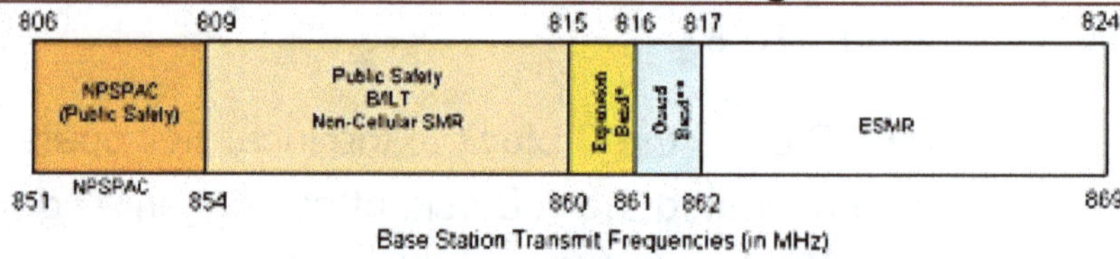

*No public safety system will be required to remain in or relocate to the Expansion Band, although they may do so if they choose.
**No public safety or CII licensee may be involuntarily relocated to occupy the Guard Band.

Allocation	Bandwidth (MHz)	Frequencies
NPSPAC (Public Safety)	6 MHz	806–809 MHz, 851-864 MHz
Public Safety/ non-cellular SMR	12 MHz	809-815 MHz, 854-860 MHz
Expansion Band	2 MHz	815-815 MHz, 860-861 MHz
Guard Band	2 MHz	816-817 MHz, 661-862 MHz
ESMR	14 MHz	817-824 MHz, 862-869 MHz
Cellular	50 MHz	824-849 MHz, 869-894 MHz

Source: FCC and Summit Ridge Group, LLC analysis.

C. Personal Communications Service (PCS): 1.8 GHz and 1.9 GHz

The PCS spectrum is divided into narrowband PCS and broadband PCS licenses

In the 1990s after the initial frequency bands used for mobile services (806 to 902 MHz) were exhausted, the FCC allocated a higher spectrum range to cellular phone services called Personal Communication Services (PCS), which is solely used to transmit digital signals. PCS is a part of the spectrum that contains the 130 MHz band from 1850 to 1915 MHz (for uplink) and 1930 to 1995 MHz (for downlink). It is divided into seven frequency blocks (A-G), each of which has a bandwidth between 10 and 30 MHz. It supports a variety of technologies including GSM, 3G and 4G technologies. The PCS spectrum is also divided into narrowband PCS and broadband PCS licenses, where narrowband PCS spectrum is allocated to relatively simple two-way text-based services while broadband PCS spectrum is for more complex two-way wireless data and voice services (mobile and fixed radio). The PCS spectrum was auctioned across the country in geographical units called Major Trading Areas (MTA). The PCS Band plan is illustrated in Exhibit 9-6 below. Note that the initial C Block licenses (originally 30 MHz each) were split into multiple licenses: C-1 and C-2 with 15 MHz each; and C-3, C-4, and C-5 with 10 MHz each.

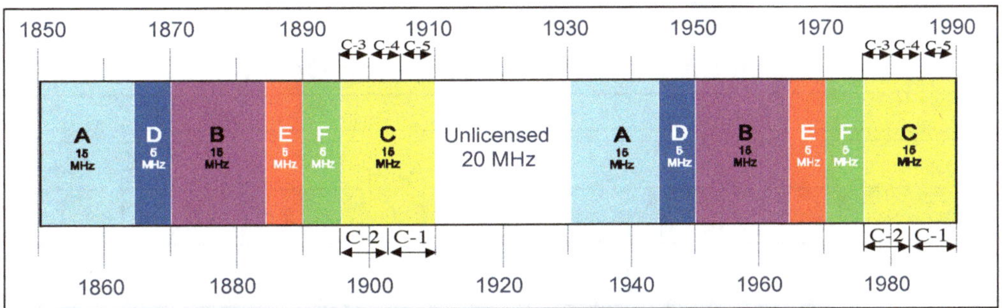

Channel Block	Frequencies (MHz)	Bandwidth (MHz)	Pairing
A	1850-1865, 1930-1945	30	2 x 15 MHz
B	1870-1885, 1950-1965	30	2 x 15 MHz
C	1895-1910, 1975-1990	30	2 x 15 MHz
D	1865-1870, 1945-1950	10	2 x 5 MHz
E	1885-1890, 1965-1970	10	2 x 5 MHz
F	1890-1895, 1970-1975	10	2 x 5 MHz
Duplex Gap	1910-1930	20	NA

Source: FCC and Summit Ridge Group, LLC analysis.

The FCC introduced MSS licenses in 1997, which currently occupy a total of 90 MHz of spectrum in the L-band and S-band

D. L-Band (1.6 GHz) and S-Band (2.0 GHz)

Mobile Satellite Services (MSS) operate in the 1.6 GHz and 2.0 GHz frequency bands, known as the L-band and S-band respectively. The FCC introduced MSS licenses in 1997, covering a total of 90 MHz of spectrum. As discussed in Section V, the FCC is seeking to free this spectrum for dual satellite/terrestrial authorization.

E. Advanced Wireless Spectrum (AWS): 1.7 GHz and 2.1 GHz

The AWS spectrum is intended for fixed and mobile voice and data services. Current service rules for these bands can be found at 47 C.F.R Part 27. The AWS spectrum contains multiple sections. A band plan for AWS-1 spectrum and for the 2 GHz AWS 2, 3 and 4 spectrum as a whole is illustrated in Exhibits 9-7 and 9-8 respectively below.

Exhibit 9-7: AWS 1 Band Plan Detail

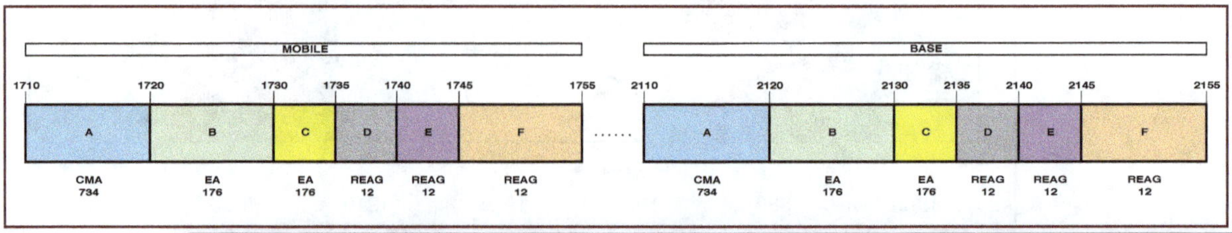

Block	Frequencies (MHz)	Bandwidth	Paring	Area	Licenses
A	1710-1720, 2110-2120	20 MHz	2 x 10 MHz	CMA	734
B	1720-1730, 2120-2130	20 MHz	2 x 10 MHz	EA	176
C	1730-1735, 2130-2135	10 MHz	2 x 5 MHz	EA	176
D	1735-1740, 2135-2140	10 MHz	2 x 5 MHz	REAG	12
E	1740-1745, 2140-2145	10 MHz	2 x 5 MHz	REAG	12
F	1745-1755, 2145-2155	20 MHz	2 x 10 MHz	REAG	12

Source: FCC[107] and Summit Ridge Group, LLC analysis.

[107] Available at http://wireless.fcc.gov/services/aws/data/awsbandplan.pdf.

Exhibit 9-8: 2 GHz AWS-2,3 & 4 Band Plan

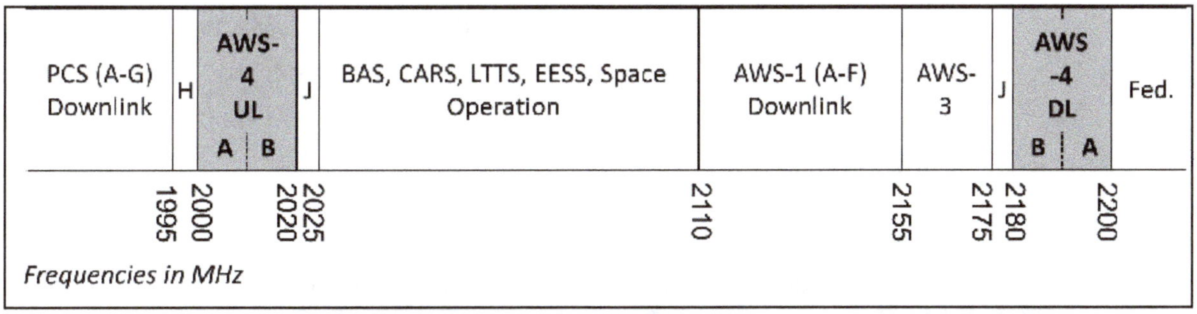

Application	Frequencies (MHz)	Bandwidth	Pairing
AWS- 2 H Block	1915-1920 1995-2000	10 MHz	2 x 10 MHz
AWS- 2 J Block	2020-2025, 2175-2180	10 MHz	2 x 5 MHz
AWS- 3	2155 – 2175	20 MHZ	NPRM pending @ FCC
AWS-4 A Block	2000-2010, 2190-2200	20 MHz	2 x 10 MHz
AWS-4 B Block	2010-2020, 2180-2190	20 MHz	2 x 10 MHz

Source: FCC and Summit Ridge Group, LLC analysis.

Federal government agencies currently use much of the AWS-1 uplink spectrum

AWS-1 spectrum contains the frequency ranges 1710 to 1755 MHz for uplink and 2110 to 2155 MHz for downlink. It is divided into six frequency blocks (A-F). Federal government agencies currently use much of the uplink spectrum and will need to be moved to allow for commercial uses. There is much uncertainty over the cost and timing of this move. The licenses for Block A are issued for the 734 Cellular Market Areas, Blocks B and C are issued for the 176 Economic Areas, while blocks D, E and F are issued for each of the 12 Regional Economic Groups in the U.S. There have been two AWS auctions: FCC Auction No. 66 in 2006 and Auction No. 78 in 2008.

AWS-2 spectrum contains the frequency ranges 1915 to 1920 MHz and 1995 to 2000 MHz (H block – paired), 2020 to 2025 MHz and 2175 to 2180 MHz (J-block – paired). Proposals for this spectrum are pending before the FCC. The FCC's 2004 Notice of Proposed Rulemaking for this spectrum is also still pending. The AWS-2 H block has power limits on the 1917 to 1920 portion, limiting its use.

AWS-3 spectrum runs from 2155 to 2175 MHz. Proposals for this spectrum are currently pending before the FCC. Many expect it will ultimately be paired with 1755 – 1780 MHz spectrum. The pairs would be equal-sized if the J Block (2175 to 2180 MHz) were added to the AWS-3 spectrum.

AWS-4 spectrum runs from 2000 to 2020 and 2180 to 2200 MHz. This spectrum is currently licensed to DISH. While it was originally subject to ATC requirements (see Section IV.E.2 above for background on ATC), the FCC recently eliminated the ATC requirements for this band, allowing it to be used with terrestrial-only handsets. This significantly increased the value of the spectrum by lowering the cost of the handsets.

SiriusXM Satellite Radio controls 25 MHz between the two WCS blocks for its satellite radio service

F. Wireless Communication Service (WCS): 2.3 GHz

The WCS band consists of 30 MHz in the frequency range 2305 to 2320 MHz and 2345 to 2360 MHz. The FCC intends for the WCS band to be used for wireless broadband, by dividing it into two 10 MHz blocs (paired 5 MHz blocks) and two unpaired 5 MHz blocks. The satellite radio provider SiriusXM Satellite Radio controls the 25 MHz between the two blocks for its two separate satellite networks. An illustrated description of the WCS band is in Exhibit 9-9 below.

Exhibit 9-9: WCS Band Plan

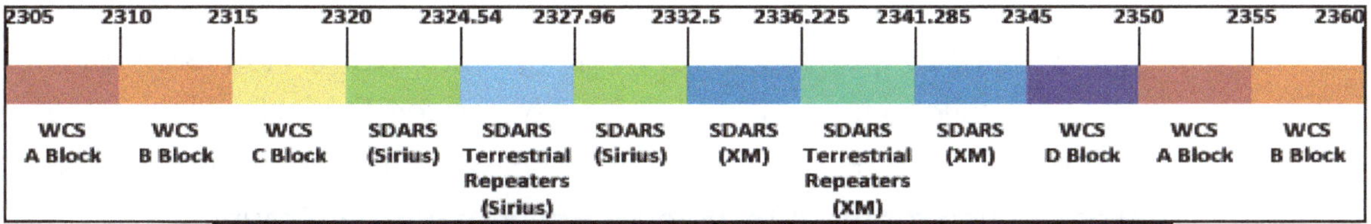

Application	Frequency (MHz)	Bandwidth (MHz)	Pairing
WCS – A Block	2305-2310, 2350-2355	10 MHz	2 x 5 MHz
WCS – B Block	2310-2315, 2355-2360	10 MHz	2 x 5 MHz
WCS – C Block	2315-2320	5 MHz	Unpaired
SDARS (Sirius)	2320-2324.54, 2327.96-2332.5	9.08 MHz	2 x 4.54 MHz
SDARS Repeaters (Sirius)	2324.54-2327.96	3.42 MHz	Unpaired
SDARS (XM)	2332.5-2336.225, 2342.285-2345	7.430 MHz	3.725 MHz 3.715 MHz
SDARD Repeaters (XM)	2336.225-2341.285	5.060 MHz	Unpaired
WCS – D Block	2345-2350	5 MHz	Unpaired

Source: FCC and Summit Ridge Group, LLC analysis.

In October 2012, the FCC adopted rules to enable WCS and satellite radio to operate without interference

One significant technical issue that the WCS band faces is interference with DARS (Digital Audio Radio Service) spectrum, which abuts the frequency ranges used for WCS.

AT&T and SiriusXM entered into a voluntary agreement in the Summer of 2012 to solve this issue. On October 17, 2012 the FCC effectively approved this agreement by adopting its key terms as service rules to enable WCS and satellite radio to operate without interference.[108] These service rules allow 20 MHz to be used for mobile broadband service and 10 MHz for fixed broadband service. The FCC has indicated that some portion of the fixed allocation

[108] 47 C.F.R. Part 27. See also, AT&T Public Policy Blog, FCC Approves AT&T, Sirius XM WCS Spectrum Band Proposal, October 17, 2013, accessed on May 15, 2013, available at http://attpublicpolicy.com/fcc/fcc-approves-att-sirius-xm-wcs-spectrum-band-proposal.

may be used for mobile services in the future.

WCS licenses are issued via auction, but only one has taken place-- FCC Auction No. 14 in 1997. The A and the B blocks are issued for the 52 Major Economic Areas (MEAs) while the C and D blocks were issued for the 12 Regional Economic Area Groupings (REAGs). WCS spectrum use is subject to FCC service rules under 47 C.F.R. Part 27.

A majority of the BRS spectrum remains unused

G. BRS/EBS (2.5 GHz)

A majority of the BRS spectrum remains unused. Accredited educational entities control licenses for the more than two thirds of the BRS spectrum allocated for Educational Broadband Services (EBS). EBS licensees have the right to this frequency range indefinitely, but do not need to use all of it for educational services under the terms of their licenses. They lease a significant amount of its spectrum to private wireless companies including Clearwire and Sprint/Nextel. Clearwire controls approximately 133 MHz of the 194 MHz of BRS/EBS spectrum. (Note: to date, Sprint Nextel and Clearwire have agreed to merge, but have not completed the transaction. DISH has challenged the deal with a competing bid.) The current EBS/BRS Band plan is illustrated in Exhibit 9-10 below.

Exhibit 9-10: EBS/BRS Band Plan

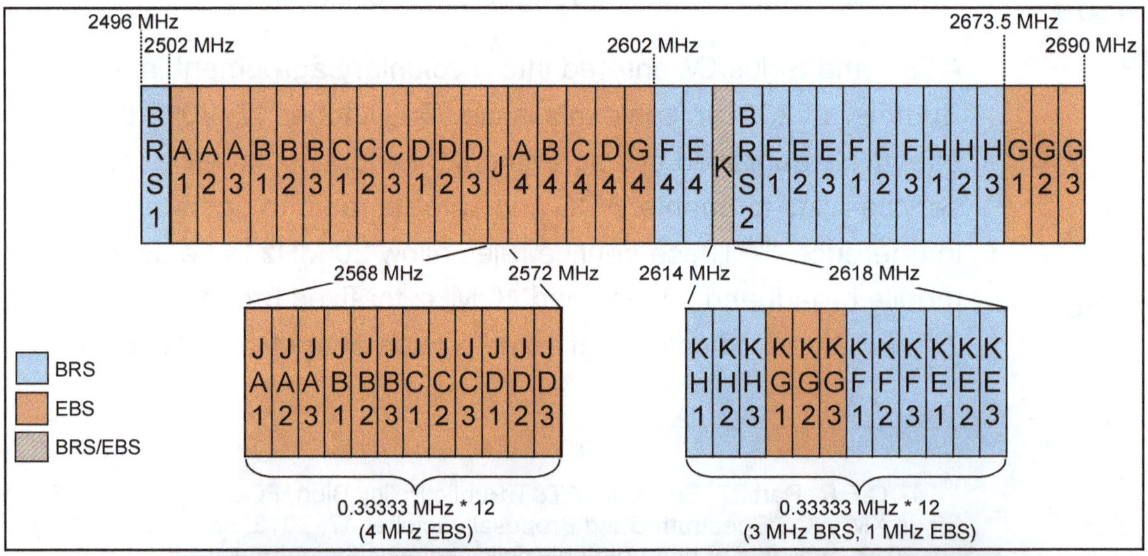

Type	Frequency (MHz)	Amount	Type

BRS	2496-2502	6 MHz	Unpaired
EBS	2502-2602	100 MHz	Unpaired
BRS	2602-2615	13 MHz	Unpaired
EBS	2615-2616	1 MHz	Unpaired
BRS	2616-2673.5	57.7 MHz	Unpaired
EBS	2673.5-2690	16.5 MHz	Unpaired

Source: FCC and Summit Ridge Group, LLC analysis.

X. Radio and Television Broadcasting Band Descriptions

The spectrum allocated to radio (AM and FM) and television (VHF and UHF) services lies between 540 kHz and 698 MHz. For both radio and television broadcasting, these frequency bands are divided into different stations that operate at specified maximum power levels and are spaced apart appropriately for the purpose of preventing interference with different stations. In recent years, radio and television broadcasting have transitioned from traditional analog transmission to digital transmission, which uses binary signals that are significantly more efficient.

Radio broadcasting has maintained its analog broadcasts alongside digital broadcasting while television broadcasting eliminated analog signals in 2009

Radio broadcasting has maintained its analog broadcasts alongside digital broadcasting while television broadcasting eliminated analog signals in 2009. This allowed stations to add additional services or reduce their spectrum needs. Television licenses are 6 MHz each. Currently 294 MHz is allocated to television broadcasting, but the FCC is seeking to reduce this by 120 MHz via the expected incentive auction process to provide more spectrum for mobile broadband. Summit Ridge Group expects that the FCC will ultimately decide to eliminate channels 31 to 51 and repack the remaining broadcasters in the lower 30 channels. See Section IV.D and V.A above for additional information about the FCC's proposed television incentive auction.

A. AM Radio: 540 kHz to 1700 kHz

AM (amplitude modulation) radio, varies the amplitude (height) of the radio waves while keeping the wavelength or frequency constant to transmit and receive audio signals. It is the simplest radio technology. AM Radio frequencies are issued in three categories: Clear, Regional and Local. Additionally there are four classes (A-D) of AM licenses – each station has two classes, one local and the other regional. These overlapping groupings determine the permitted power and hours of station operation. AM radio licenses typically are issued for a period of eight years and licensees are subject to rules of service under 47 C.F.R. 73 Subpart A. AM stations are allowed a maximum transmission bandwidth of 10.2 kHz.

FM radio spectrum modulation is more complex than AM radio spectrum modulation

B. FM Radio: 92.1 to 107.9 MHz

FM radio spectrum modulation is more complex than AM radio spectrum modulation. It uses frequency modulation (FM), which varies the frequency of the carrier wave (keeping the height [amplitude] of the radio wave constant) to broadcast audio signals. FM Radio stations are subject to rules of service under 47 C.F.R. Subpart B. FM stations in the U.S. are 200 kHz apart. They typically broadcast using a 15 kHz signal and a 38 kHz stereo subcarrier. This allows nearly 75 kHz of signal deviation with adjacent channel interference. There are seven types of FM radio station licenses. Their power limits are described in Exhibit 10-1 below.

Exhibit 10-1: FM Radio Licenses

Station Class	Maximum ERP	Reference HAAT in meters (ft.)	Class contour distance in kilometers
A	6 kW (7.8 dBk)	100 (328)	28
B1	25 kW (14.0 dBk)	100 (328)	39
B	50 kW (17.0 dBk)	150 (492)	52
C3	25 kW (14.0 dBk)	100 (328)	39
C2	50 kW (17.0 dBk)	150 (492)	52
C1	100 kW (20.0 dBk)	299 (981)	72
C0	100 kW (20.0 dBk)	450 (1476)	83
C	100 kW (20.0 dBk)	600 (1968)	92

Source: FCC and Summit Ridge Group, LLC Analysis.

In addition to the categories above, noncommercial educational, public safety and transportation entities may be eligible for low power (100 watt maximum) FM licenses that have a range of approximately 3.5 miles (5.6 kilometers). Low Power FM licensees are subject to rules contained in 47 C.F.R. Part 73 Subpart G.

C. VHF Television: 54 to 88 MHz and 174 to 216 MHz

VHF signal propagation is superior [to UHF], but only by a small margin

The lower portion of this band (54 to 88 MHz) includes channels 2 though 6 while channels 7 through 13 are in the upper portion (174 to 216 MHz). Due to its lower frequency range relative to UHF television, VHF signal propagation qualities provide slightly longer range. This partially accounts for why stations that carry VHF television are more valuable than their UHF counterparts.

D. UHF Television: 470 to 698 MHz

Ultra High Frequency (UHF) television broadcasting incudes channels 14 through 51 except channel 37 (608 to 614 MHz), which is allocated to radio astronomy in the United States, Canada and a few other countries. The UHF band was reduced in 2009 to the current 470 MHz to 698 MHz band by moving channels 52-69 into lower channels during the digital transition. Before this reduction the UHF television band occupied 470 to 806 MHz. The FCC had previously eliminated channels 70 to 83 (806 MHz to 890 MHz) to create the original cell phone and public safety bands. This progressive reduction of television broadcast spectrum and reallocation to mobile services is an international trend in most economically developed countries.

Progressive reduction of television broadcast spectrum and reallocation to mobile services is an international trend in most economically developed countries

Television licenses are available in three categories: Full Power, Class A and Low Power. Full power stations may broadcast in television market areas and have rights against interference in their areas. Low power stations are required to broadcast at low power so as to not interfere with either Full Power or higher priority low power broadcasters known as Class A stations and have no rights against interference from them.

Full power television broadcast licensees, both VHF and UHF, are subject to rules in 47 C.F.R. Subpart E. Class A television broadcast licensees are subject service rules in 47 C.F.R. 73 Subparts J. Low power televisions broadcasters are subject to service rules in 47 C.F.E. 73 Subpart L.

XI. Satellite Frequency Band Descriptions

Most satellites are positioned over the equator in a geostationary position so they rotate at the same rate as the earth

Most satellites are positioned over the equator in a geostationary position so that they rotate in synchronicity with the earth, thus appearing stationary from earth. To avoid interference, regulations require operators of satellites using the same frequencies to position them two degrees of longitude apart. This creates 180 possible locations for using each band of spectrum. There are exceptions, such as DBS (satellite television) in the United States, where the high power levels and small user equipment require nine degrees of spacing. To put a satellite in an orbital slot, the operators require permission from the country with rights to that orbital slot. Most countries also require satellite operators to obtain licenses to send signals to customers in that country. While the FCC generally assigns orbital slots by auction, International Telecommunication Union (ITU) rules prevent countries from auctioning orbital slots that it considers international – those serving multiple countries. Despite the ban on auctioning licenses for using such slots, they are effectively sold on the secondary market.

C-band is used primarily for satellite transmission and is desirable in tropical areas due to its resistance to rain fade

A minority of satellites are not in geostationary orbit, but in Middle Earth Orbit (MEO) or Low Earth Orbits (LEO). These satellites move faster than the earth so a fleet of them is needed for constant coverage of a given area. Among these are the Iridium and Globalstar satellite fleets, each with dozens of satellites, using a low earth orbit to minimize latency (delay) on calls. The orbits and spectrum for non-geostationary satellites are coordinated through the ITU. However these service providers must negotiate rights to sell their services and the royalty rights for their customers to use their equipment in each country that they operate. This can be a daunting task.

With increased congestion in the lower frequencies, satellite service provides have increasingly turned to higher frequencies - enabled by higher power satellites

Satellite communication is less sensitive to differences in frequency than terrestrial communication because the power levels from the satellite are so low that a constant line of sight is generally necessary regardless of spectrum, and there are no ionospheric bouncing issues to contend with. However, as with terrestrial communication, lower frequencies are blocked less – in the case of satellites primarily by rain and light foliage. But with increased congestion in the lower frequencies, satellite service providers have increasingly turned to higher frequencies – enabled by higher power satellites that can penetrate weather related blockage. Until the late 1980s, C-band was the dominant satellite frequency band; the Ku-band and Ka-band were considered unusable. Nearly a decade later, higher-powered satellites that allowed Ku-band to become popular. Now, Ka-band is becoming increasingly popular as improved solar panel technologies make construction of even higher powered satellites possible. Support is growing for satellites in the V-band. If such a satellite could be built, it would have significantly greater data capacity, which could make satellite broadband more competitive with terrestrial alternatives. Descriptions of satellite bands follow. Note that satellite frequency bands deviate slightly from the traditional IEE Band descriptions in Exhibit 2-2 in Section II.

A. L-Band: 1.4 GHz and 1.6 GHz

Satellite broadcasters are allocated the 1482.352 to 1490.624 MHz and 1525 to 1646.5 MHz bands. The former is used primarily for satellite radio and related services; the latter for satellite telephony (MSS) services. The middle part of the band is reserved primarily for radio astronomy, military and global positioning systems such as the European Galileo and the Russian GLONASS systems. The L-band frequency plan is illustrated in Exhibit 11-1 below.

Exhibit 11-1: L-Band Frequency Plan

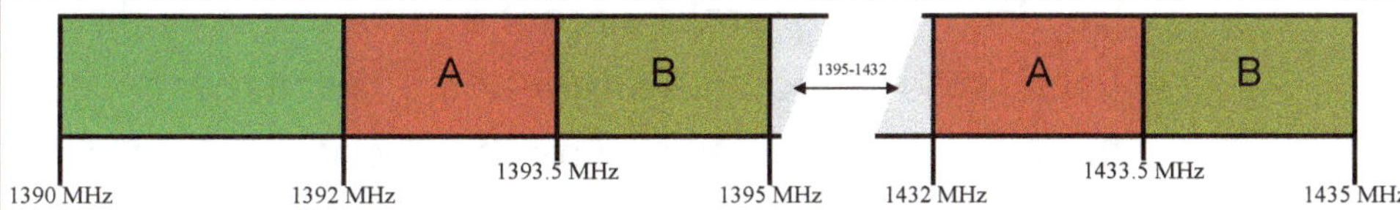

Block	Frequencies (MHz)	Bandwidth	Pairing	Geo. Area Type	No. of Licenses
	1390-1392	2 MHz	Unpaired	MEA	52
A	1392-1393.5/1432-1433.5	3 MHz	2 x 1.5 MHz	EAG	6
B	1393.5-1395/1433.5-1435	3 MHz	2 x 1.5 MHz	EAG	6

Block	Frequency (MHz)	Bandwidth	Pairing	Area	Licenses
	1390-1392	2 MHz	Unpaired	MEA	52
A	1392-1393, 1432-1433.5	3 MHz	2 x 1.5	EAG	6
B	1393.5-1395, 1433.5-1435	3 MHz	2 x 1.5	EAG	6

Source: FCC and Summit Ridge Group, LLC analysis.

Many countries are considering reallocation of the lower portion of the C-band receive band for terrestrial Wi-Fi use

B. S-band: 2.0 to 2.7 GHz

The S-band is used primarily for satellite radio services (2.31 to 2.36 GHz) as well as mobile satellite services (portions between 2.0 and 2.2 GHz) and the AWS mobile services. The 2.4 to 2.483 GHz segment is allocated for unlicensed use such as Wi-Fi, blue tooth and cordless telephones. Non-radio equipment such as microwave ovens use this band and, unlike radios, utilize the entire band while operating but have short distances thus mitigating interference issues.

C. C-band: 3.4 to 6.725 GHz

The standard C-band uses 3.625 to 4.2 GHz (downlink) and 5.85 to 6.425 GHz (uplink). The extended version of the C-band uses 3.4 to 3.65 GHz (downlink) and 6.425 to 6.725 GHz (uplink). The Indian National Satellite Systems have a super extended version of this band that includes 4.5 to 4.8 GHz (downlink) and 6.725 to 7.025 GHz (uplink). C-band is used primarily for satellite transmission and is desirable in tropical areas due to its resistance to rain fade. Regulators in many countries are considering reallocating the lower portion of the receive band for terrestrial Wi-Fi use. Fixed Satellite Service (FSS) operators are the main users. A disadvantage of C-band is that it requires larger antennas to receive signals, making them harder to install and more expensive. This limits widespread C-band use by satellite television consumers.

D. Ku-band: 10.7 to 18 GHz

Ku-band is used primarily in satellite communications. Most commercial satellite television signals in the United States and Europe are transmitted in Ku-band. Although not as resistant to rain fade as C-band, its higher frequency allows consumers to use much smaller reception antennas. The Ku-band allocation is more complex than the C-band. In ITU Region 2 (the Americas), this band is used for FSS (11.7 to 12.2 GHz for downlink and to 14.0 to 14.5 GHz for uplink) DBS (12.2 to 12.7 GHz for downlink and 17.3 to 17.8 GHz for uplink) services.

Most commercial satellite television signals in the United States and Europe are transmitted in Ku-band

In ITU region 1 (Europe and Africa), the frequencies are slightly different: FSS uses 10.7 to 11.7 GHz for downlink and 14.0 to 14.8 GHz for uplink while DBS uses 11.7 to 12.5 GHz for downlink and 17.3 to 18.1 GHz for uplink. In Australia (ITU Region 3), the Ka-band downlink is 12.25 to 12.75 Ghz and uplink is from 14.0 to 14.5 GHz.

E. X-Band: 7 to 11.2 GHz

Military communication satellites primarily use X-band at 7.25 to 7.75 GHz (downlink) and 7.9 to 8.4 GHz (uplink). It is also used for radar applications, deep space applications and amateur radio. Some motion detectors use the 10.525 GHz frequency.

F. Ka-band: 26.5 to 40.0 GHz

The newest addition to the commercial FSS bands is the Ka-band. Ka-band satellite have historically suffered from significant weather related interference, which made it less valuable than C-band and Ku-band. However, new higher-powered satellites over the past 10 to 15 years have largely overcome this problem. This band is heavily used in state of the art satellite broadband systems. A list of Ka-band frequencies is in Exhibit 11-2 below.

Exhibit 11-2: Authorize U.S. Satellite Ka-band Frequencies

Frequency	Amount
18.3 to 18.8 GHz	500 MHz
19.7 to 20.2 GHz	500 MHz
28.35 to 28.6 GHz	250 MHz
29.25 to 30.0 GHz	750 MHz
Total	**2,000 MHz**

Source: FCC and Summit Ridge Group, LLC analysis.

The upcoming Wi-Fi standard 802.11ad will use frequency in the 60 GHz range

Local Multipoint Distribution Service (LMDS) is another technology that operates in the Ka-band at 28 to 31 GHz, transmitting signals from a single point to multiple points or a single point (LMDS is described in greater detail in Section XV.A.3). It requires line of sight between the reception and transmission antennas and a range of approximately 1.5 miles (2.4 km). In the U.S., LMDS was intended to be used for "wireless cable" but most video providers abandoned it in favor of other technologies. Today LMDS is used to backhaul mobile traffic from cell phone towers, particularly in Europe. There is some discussion of using LMDS for residential broadband in rural areas.

G. V-band: 50 GHz to 75 GHz

The V-band is used as a link between certain military satellites. Otherwise it is lightly used, mostly for research purposes. Many believe that this band will be developed for commercial satellite communications as the Ka-band becomes more congested. The upcoming Wi-Fi standard 802.11ad will use frequency in the 60 GHz range.

One advantage of high frequencies is their significantly increased capacity to carry data. Satellites in this band could increase their data capacity to the point that they are competitive with terrestrial broadband options. However, they must overcome the obstacles of blockage from atmospheric conditions. This will require the development of higher-powered satellites, which in turn require improvements in solar panel technologies for power.

XII. Government/Military Frequency Use Descriptions

A. U.S. Federal Agencies Uses Large Amounts of Spectrum

60% of spectrum is controlled primarily by federal government agencies

Almost 60 percent of available spectrum is allocated for government use in the U.S.[109] The spectrum controlled by the federal government is among the most desirable, located in all major bands. Exhibit 12-1 illustrates the spectrum bands used by the federal government.

Exhibit 12-1: U.S. Government Spectrum Usage

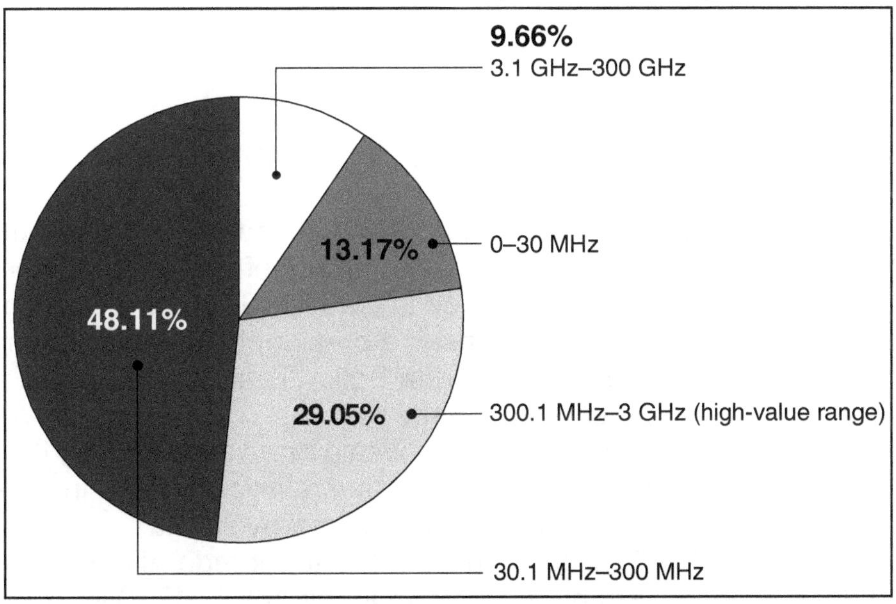

Source: GAO analysis of NTIA GMF spectrum assignment data, September 7, 2010.

B. Defense Is Largest Government User

[109] PCAST Report, 8.

The largest portion of spectrum used by the Federal government is for national defense, followed by law enforcement. Exhibit 12-2 below illustrates the use of spectrum by federal agencies.

Exhibit 12-2: U.S. Government Agencies with the Most Spectrum

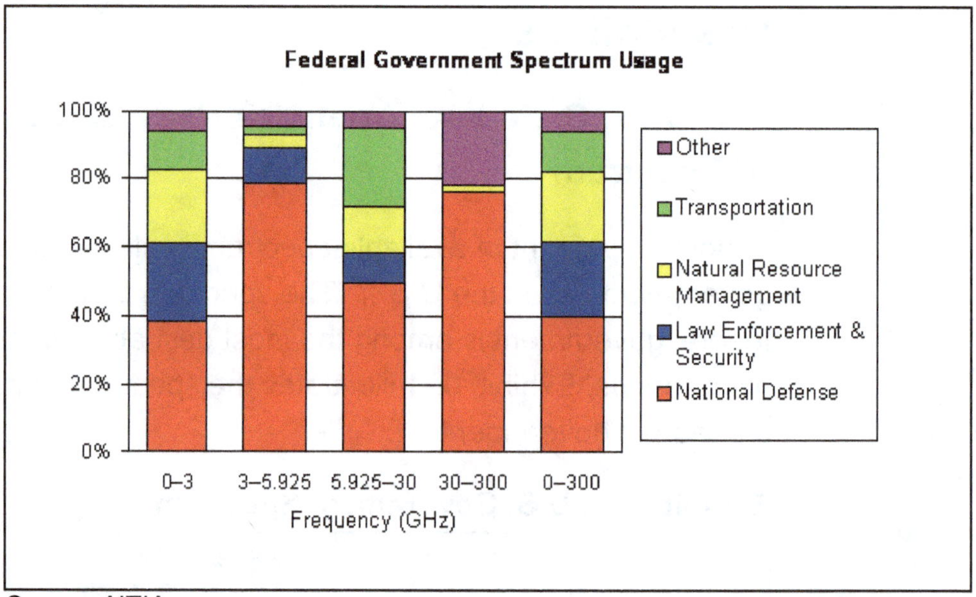

Source: NTIA.

C. Government Spectrum Trends

As mentioned previously, the U.S. federal government agencies have been heavily criticized for not efficiently using their spectrum. This criticism culminated in the PCAST report. On June 14' 2013, the While House released a presidential memorandum, ordering the creation of a Spectrum Policy Team ("Spectrum Team"). The Spectrum Team is ordered to;

1) Produce a report within one year detailing the FCC and NTIA's spectrum sharing initiatives and include recommendations for improvement.
2) Facilitate current plans to for federal agencies sharing or returning spectrum in the 1697 - 1710 MHz band, the 1755 – 1850 MHz band, the 5350 – 5470 band and the 5850 – 5925 MHz band. They are also directed to facilitate sharing of other spectrum under 6 GHz.
3) Publish and inventory of test facilities for sharing spectrum and a standards for sharing within 6 months.
4) Evaluate each federal agency's use of its spectrum and each agency is directed to submit its own assessment.

Each agency that uses spectrum between 400 MHz and 6 GHz is directed to only keep the minimum spectrum needed for its mission.

5) Expansion of testing of federal agency's use of their spectrum.
6) Development of performance criteria for receivers using shared spectrum.
7) Propose market-based incentives for agencies to encourage them to relinquish or share their spectrum.
8) Encourages the Spectrum Team to identify ways to redeploy any returned spectrum for commercial broadband use.

It's unclear if this initiative will actually result in federal agencies giving up or sharing a significant amount of spectrum. However, it's clear that there is a trend towards putting increased pressure on them to do so.

XIII. Very High Frequency (VHF) Services

VHF frequencies work well for short distances – somewhat longer than the range of sight

Frequencies in the VHF frequency band work well for short distances – somewhat longer than the range of sight. Unlike the High Frequency spectrum discussed above, VHF transmissions do not extend their range by "bouncing" off the ionosphere. The VHF spectrum band includes a number of services. Some common VHF applications are discussed below.

A. Lower Paging: 35, 43, 152 & 158/159 and 454 MHz

The lower paging band occupies several blocks of frequencies. The 35, 43, 152 and 158 MHz blocks are not paired. However, part of the 152 block is paired with 159 MHz and 454 MHz band is paired within the 454 MHz block. The paging industry has largely been in decline as mobile phones with greater functionality have dropped in price. An illustration of the lower paging band plan is shown in Exhibit 13-1 below.

Exhibit 13-1: Lower Paging Band Plan

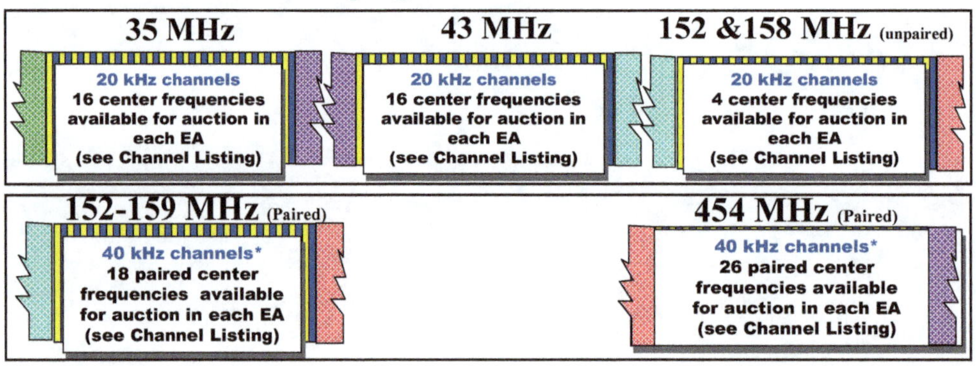

Source FCC.

B. Radio Control Service: 72 to 73 MHz and 75.4 to 76 MHz

Located in the 72.0 to 73.0 MHz and 75.4 to 76.0 MHz spectrum ranges, radio control spectrum is commonly used for short-distance, one-way communications for operating devices that have an on/off switch located at a place distant from the operator. This includes devices such as model aircrafts and cars. Service rules for this frequency band are located at 47 C.F.R. Part 95.

C. VHF Public Coast: 150.8 to 162.025 MHz

Channel 16 of the VHF Pubic Coast Band is the international calling and distress channel

This band is divided into two parts, a marine mobile services portion and a land mobile services portion. Marine VHF is installed on virtually all large ships as well as most small craft. It is used for safety purposes including collision avoidance. Channel 16 of the VHF Public Coast Band is the international calling and distress channel. Power levels range from 1 to 25 watts, giving a maximum range of 60 nautical miles with an antenna mounted on a tall ship or 5 nautical miles when using small antennas mounted at sea level.

On land, the 151 to 154 MHz is used for short-distance, two-way communication, usually via small handheld devices. The spectrum is divided into five channels of 11.25 or 20 kHz each. An illustration of the band plan for the VHF Public Coast band is shown in Exhibit 13-2 below. Service rules for this frequency band are located at 47 C.F.R. Part 95.

Exhibit 13-2: VHF Public Coast Band Plan

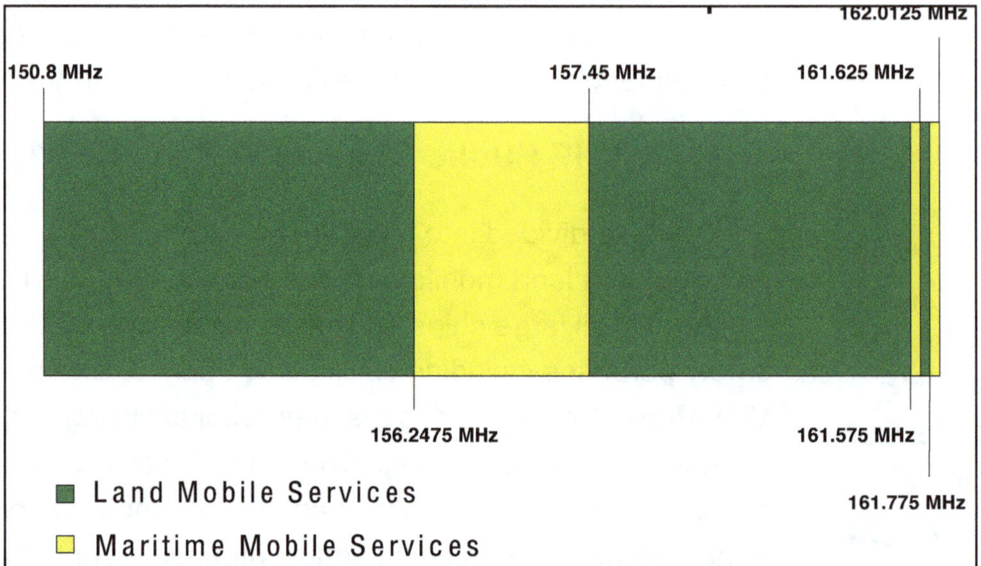

Note: Public Safety Set-Aside Chart for Inland VPCs

Application	Frequencies (MHz)	Bandwidth	Type
Land Mobile Services	150.8-156.2475	5.4475 MHz	Unpaired
Maritime Mobile Services	156.2475-157.45	1.1925 MHz	Unpaired
Land Mobile Services	157.45-161.575	4.125 MHz	Unpaired
Maritime Mobile Services	161.575-161.625	0.050 MHz	Unpaired
Land Mobile Services	161.625-161.775	0.150 MHz	Unpaired
Maritime Mobile Services	161.775-162.0125	0.2375 MHz	Unpaired

Source: FCC and Summit Ridge Group, LLC analysis.

BETRA frequencies are allocated to provide basic telephone service... where traditional wireline service is not feasible

Multi-use Radio Service

D. Basic Exchange Telephone Radio Service (BETRS)

Operated in the 152 to 159 MHz, 454 to 460 MHz, 816 to 820 MHz, and 861 to 865 MHz spectrum bands. Note: the latter two frequency blocks for this service fall outside the VHF spectrum range and into the UHF spectrum discussed in Section III.E above. The frequency blocks are broken down into 94 channels of 20 kHz each. This group of frequencies is allocated for basic telephone service in remote locations where traditional wireline service is not feasible. This spectrum is licensed on a site-by-site basis and is available to buy or lease on the secondary market. A similar service, called the Rural Radiotelephone Service also operates in the 152 to 150 MHZ and the 454 to 460 MHz bands. Service rules are located at 47 C.F.R. Part 22.

E. Automated Marine Telecommunication System (AMTS)

AMTS use is limited to 12 nautical miles off the Unites States coast

The automated marine telecommunication system is a commercial system that offers both voice and data to marine customers in coastal areas and inland waterways. Use is limited to 12 nautical miles off the United States coast. AMTS spectrum may also be used in Land Mobile Radio systems subject to interference restrictions with the adjacent television channels 10 and 13 or the Air Force Space Surveillance System operating at 216.880 MHz to 217.080 MHz. It consists of 20 channels of 25 kHz each. Users with geographical area licenses can also use 12.5 kHz or 6.25 kHz channels to increase the number of channels. An illustration of the AMTS band plan is shown in Exhibit 13-3 below.

Exhibit 13-3: Automated Marine Telecommunication System

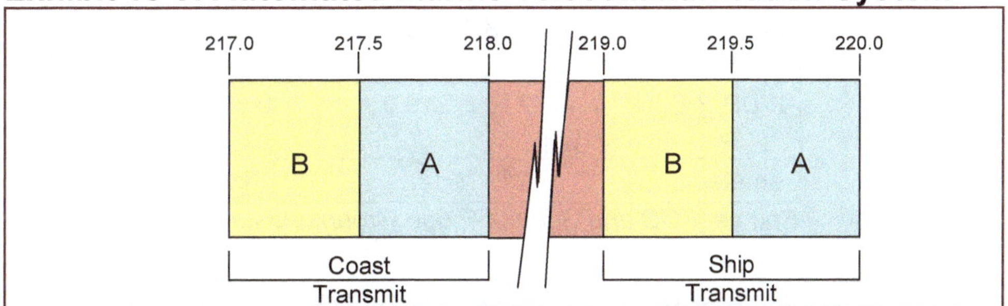

Channel Block	Frequency Bands	Bandwidth	Pairing	Channels	Licenses
A	217.5-218.0, 219.5-220.0	1 MHz	2 x 0.5 MHz	20	9
B	217.0-217.5, 219.0, 219.5	1 MHz	2 x 0.5 MHz	20	1

Source: FCC and Summit Ridge Group, LLC analysis.

F. Former Interactive Video and Data Service: 218-219 MHz

The FCC is evaluating the future use of the spectrum located at 218 to 219 MHz. Current rules can be found at 47 C.F.R. Part 95. The FCC has proposed moving the service rules to Part 27 in 2010.

This would move it from the "Personal Radio Services" category to "Miscellaneous Radio Service" category proving additional flexibility for users. But, this proposal is still pending. The spectrum is licensed in two channel blocks (A and B) in each of the 734 Cellular Market Areas (CMAs) in the United States. The FCC auctions this spectrum, but has only had one auction, FCC Auction No. 2, in 1994. This spectrum is also available on the secondary market. The band plan for this spectrum is illustrated in Exhibit XIII-4 below.

Exhibit 13-4: 218-219 MHz Band Plan

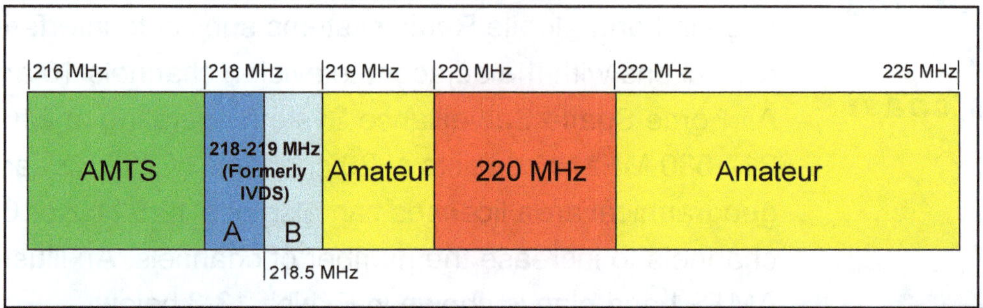

Application	Frequency	Bandwidth	License Type
AMTS	216-218 MHz	2 MHz	
IVDS A Block (former)	218.0-218.5	0.5MHz	CMA – 734 licenses nationwide
IVDS B Bock (former)	218.5-219.0	0.5MHz	CMA – 734 licenses nationwide
Amateur	219 MHz	1 MHz	

Source: FCC and Summit Ridge Group, LLC analysis.

XIV. Ultra High Frequency (UHF) Services

Frequencies in the UHF band range from ten centimeters to one meter in length. UHF radio waves do not bounce off the ionosphere as they are generally blocked by hills and large buildings. Due to their relatively short wavelength, consumer equipment using this frequency band can utilize relatively small antennas. The lower end of the band has greater distance and building penetration capabilities than the higher end.

A. Personal Locator Beacons (PLBs): 406 MHz

Personal Locator Beacons (PLBs) operate in the 406 MHz spectrum band and transmit personalized distress signals primarily for search and rescue missions

Personal Locator Beacons (PLBs) operate in the 406 MHz spectrum band and transmit personalized distress signals. These signals are designed to aid in search and rescue missions. Backwoods skiers often use PLBs so they can be found in the event they are covered in an avalanche.

B. Family Radio Service (FRS)/General Mobile Radio Service (GMRS): 462-467 MHz

The spectrum band from 462 to 467 MHz is used for short-distance two-way communication, usually via small handheld devices. GMRS is divided into 23 channels of 25 kHz each. GMRS devices are limited to 5 watts of effective radiated power (ERP) and generally have a range of five to twenty-five miles. Using FRS devices does not require a license, but they are limited to .5 watts of ERP and consequently have a much shorter range. Their use is regulated under 47 C.F.R. Part 95.

C. Specialized Mobile Radio: 800 MHz

Specialized Mobile Radio (SMR) is typically used for mobile business radio such as taxi dispatch, construction companies and other business. The radios can be mounted in a vehicle or are hand-held devices. Typically a business using SMR would pay a fee to a service provider who maintains a repeater to ensure coverage of a certain area for multiple customers. The spectrum band plan for specialized mobile radio is illustrated in Exhibit 14-1 below.

Exhibit 14-1: Specialized Mobile Radio Band Plan

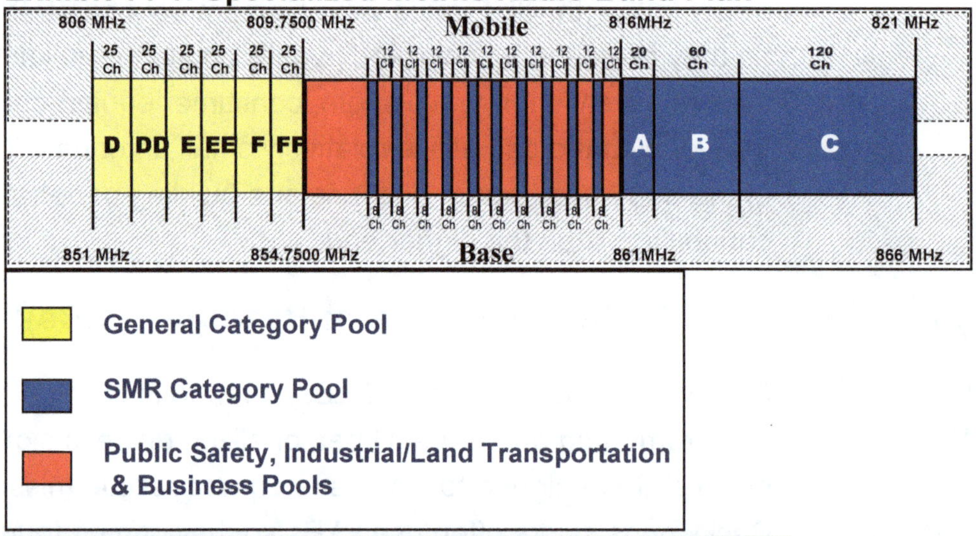

Band	Frequencies	Bandwidth	Pairing
General Category Pool	806-809.75, 851-854.75	7.5 MHz	2 x 3.75 MHz
SMR Category Pool/Public Safety (alternating)	809.75-816, 854.75-861	12.5 MHz	2 x 6.25 MHz
SMR Category Pool	816-821, 861-866	10 MHz	2 x 5 MHz

Source: FCC and Summit Ridge Group, LLC analysis.

D. Air Ground Service: 800 MHz

Air ground services are used to provide commercial telephone service to pilots and passengers aboard commercial aircraft. There are ten blocks of 6 kHz, and six control ranges of 3.2 kHz in the 849 to 851 MHz range (uplink from the ground) and in the 894 to 896 MHz (downlink from the plane). The band plan for air ground services is illustrated in Exhibit 14-2 below. The primary operator in the band was Verizon Airfone which had limited commercial success due to high consumer costs and poor call quality. Verizon Airfone was sold to Jet Blue in 2008. General Aviation users (non-commercial planes) are allocated frequencies at 454 and 459 MHz for plane to ground phone calls.

Exhibit 14-2: Air Ground Radio Telephone Service Plan (800 MHz)

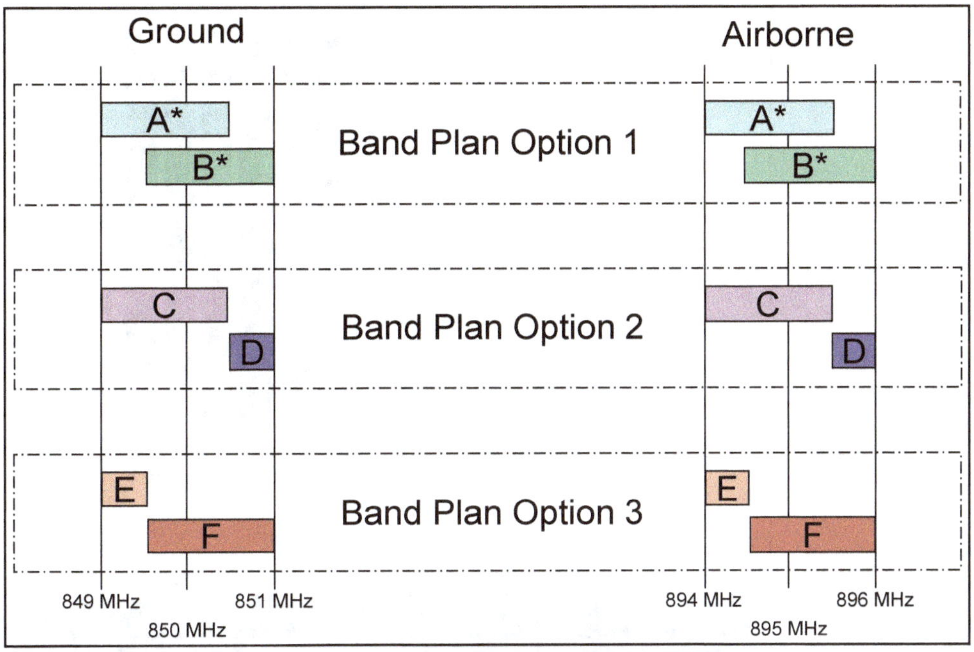

Location	Frequency	Bandwidth
Ground	849-851	2 MHz
Airborne	894-896	2 MHz

Source: FCC and Summit Ridge Group, LLC analysis.

E. Narrowband Personal Communication Service (PCS)

Narrowband PCS is located in the ranges of spectrum at 901 to 902 MHz, 930 to 931 MHz, and 940 to 941 MHz. This spectrum is typically used for two-way paging and telemetry such as off-site utility monitoring. The FCC service rules for Narrowband PCS are located at 47 C.F.R Part 24. The band plan for the PCS band is illustrated in Exhibit 14-3 below.

Exhibit 14-3: Narrowband Personal Communications Service (PCS)

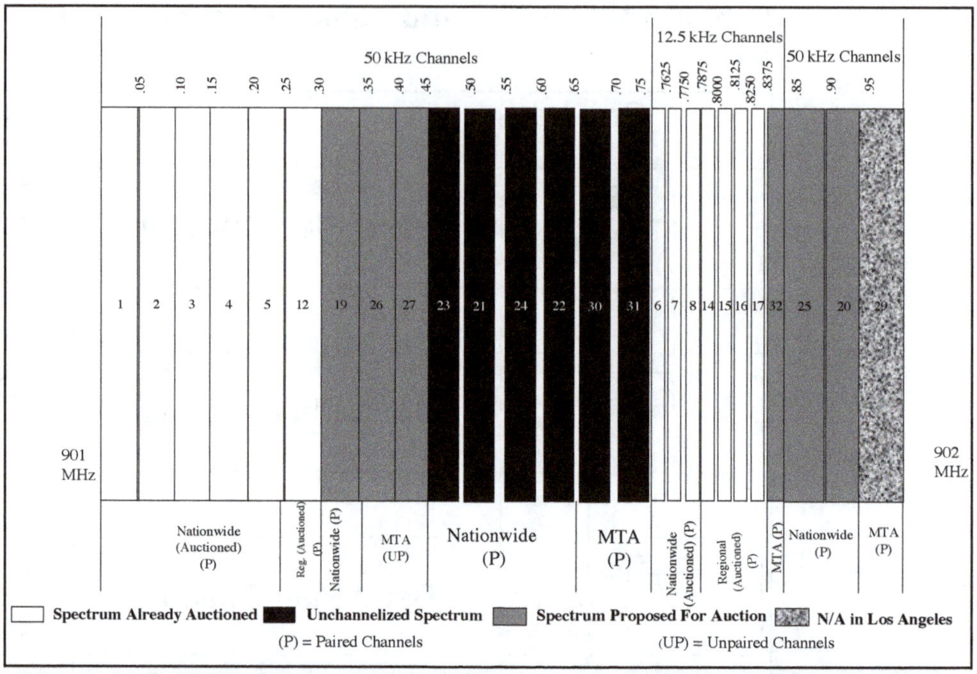

Status	Frequency	Bandwidth	Type
Nationwide (Auctioned)	901.0-901.25	250 kHz	5 channels, 50 kHz each. Paired
REG (auctioned)	901.25-901.30	50 kHz	1 channel, 50 kHz, paired
MTA (unauctioned)	901.30-901-35	50 kHz	1 channel, 50 kHz, paired
MTA	901.35-901.45	100 kHz	2 channels, 50 kHz each
Nationwide	901.45-901.65	200 kHz	4 channels, 50 kHz each, paired
MTA	901.65-901.75	100 kHz	2 channels, 50 kHz each, paired
Nationwide (unauctioned)	901.750-901.7875	37.5 kHz	3 channels, 12.5 kHz each
Regional (auctioned)	901.7875-901.8375	50 kHz	4 channels, 12.5 kHz each, paired
MTA	901.8375-901.85	12.5 kHz	1 channel, 12.5 KHz
Nationwide	901.85-901.95	100 kHz	2 channels, 50 kHz each
MTA	901.95	50 kHz	1 channel, 50 kHz each

Source: FCC and Summit Ridge Group, LLC analysis.

F. Location and Monitoring Service (LMS)/Industrial Scientific and Medical (ISM): 902 to 928 MHz

Also known as the 33 centimeter band, the location and monitoring band ranges from 902 to 928 MHz. The propagation of this frequency is highly dependent on a clear line of sight. In ITU Region 2 (the Americas) it is used for ISM as well as low powered unlicensed devices such as wireless microphones as well as Wi-Fi. In Region 2, ISM also includes 2.400 to 2.4835 GHz and 5.725 to 5.875 GHz, which are also authorized for Wi-Fi.

This band was auctioned in FCC Auction 21 in 1999. The spectrum consists of three blocks: A, B and C with 6.0, 2.25 and 5.75 MHz, respectively, in each of the 176 economic areas. The portion of this band not used by location and monitoring service is allocated to amateur radio service operators. The LMS bandplan is shown below in Exhibit 14-4.

Exhibit 14-4: LMS/ISM Band Plan

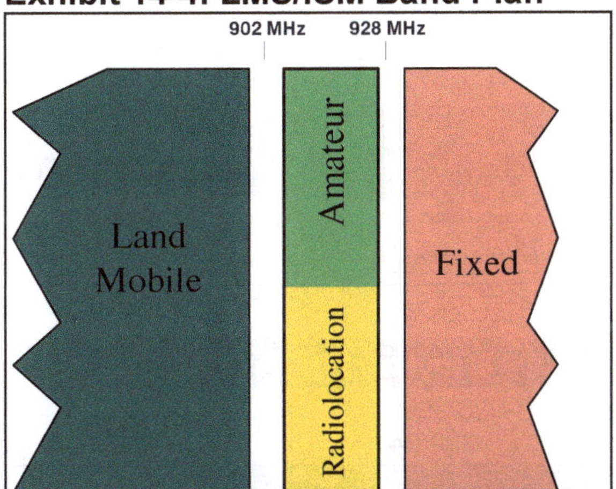

Application	Frequency	Bandwidth
Radiolocation	902-928	26 MHz
Amateur	902-928	26 MHz
ISM & formerly Wi-Fi	902-928	26 MHz
ISM & Wi-Fi	2.400 – 2.4825 GHz	82.5 MHz
ISM & Wi-Fi	5.725 – 5.875 GHz	150 MHz

Source: FCC and Summit Ridge Group, LLC analysis.

G. Multiple Address Systems: 928/959 MHz and 932/941 MHz

This Multiple Address Systems spectrum is available for terrestrial point-to-multipoint and point-to-point services, and also mobile transmission of a licensee's products or services (not including distribution of video entertainment). Users are primarily companies needing remote networks, such as energy companies using it for dedicated communications networks. Licenses were sold in FCC Auction 42 in 2001. Over 5,100 licenses were offered (twenty-nine licenses in each of the 176 economic areas) for a period of 10 years. In each market, 28 licenses were for a 25 kHz allocation while one was for 100 kHz. However, only 878 were assigned. In 2005, the FCC re-auctioned the remaining licenses in FCC Auction 59 and 2,223 of the remaining 4,226 licenses were assigned. The bandplan for Multiple Address Systems is indicated below in Exhibit 14-5.

Exhibit 14-5: Multiple Address Systems Band Plan

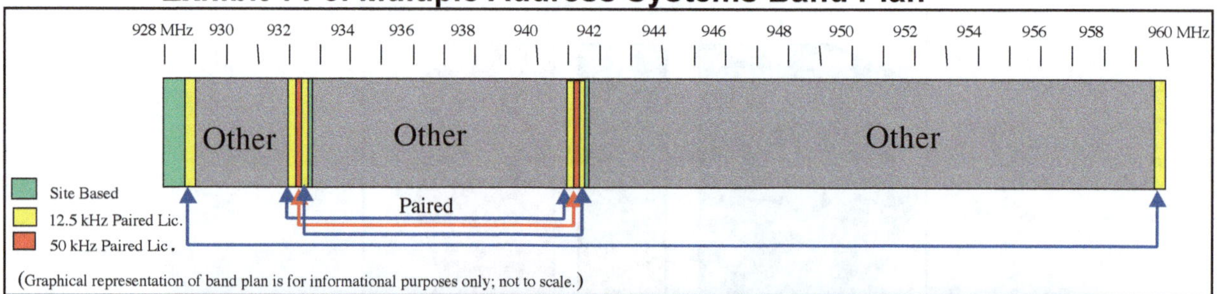

Application	Frequency	Paring	Licenses
Multiple Address Systems	928/959 and 932/941	2 x 12.5 kHz	176 (EA) 28 in each market
Multiple Address Systems	932/941	2 x 50 kHz	176 (EA) 1 in each market

Source: FCC and Summit Ridge Group, LLC analysis.

H. Upper Paging Band: 929 and 931 MHz

The upper paging band frequencies are located on either side of the narrowband PDC spectrum at 929 to 930 MHz and 931 to 932 MHz as shown in exhibit 14-6 below. The spectrum consists of 49 licenses of 20 kHz each. Twelve licenses were issued in the 929 MHz band and 37 licenses in the 931 band. The FCC issues all of these paging licenses by Major Economic Area.

Exhibit 14-6: Upper Paging Band Plan

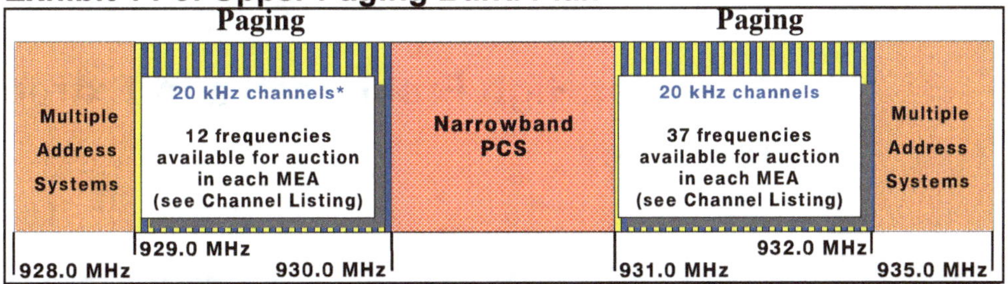

Application	Frequency (MHz)	Bandwidth	Licenses
Upper Paging	929.0-930.0	20 kHz	12 in each of 52 MEA
Upper Paging	931.0-935.0	20 kHz	37 in each of 52 MEAs

Source: FCC and Summit Ridge Group, LLC analysis.

I. Mobile Wireless Services

Commercial mobile wireless services operate in various portions of the UHF band and are discussed in detail in Section IX above.

XV. Super High Frequency and Extremely High Frequency

A. Super High Frequency (SHF): 3 GHz to 30 GHz

As mentioned in Section II.G, spectrum in this range has superior capacity to carry data on a bits per MHz basis. But propagation is largely limited to line of sight reception. Super High Frequency is used primarily by satellite communication systems (described in detail in Section XI). Exceptions include the General Wireless Communication Service (GWCS), the 24 GHz band and LMDS. In addition to its original primary band at 2.4 GHz, Wi-Fi operates in the 5 GHz band. The FCC recently announced plans to expand Wi-Fi spectrum to include parts of the 5.9 GHz band.

The SHF frequency range has superior capacity to carry data on a bits per MHz basis

1. General Wireless Communication Service

This band occupies 25 MHz-sized channels from 4660 MHz to 4685 MHz, divided into five 5 MHz blocks. This space is in the gap between the satellite C-band uplink and downlink spectrum (the "duplex gap"). The FCC scheduled it for auction in Auction 30 in May 1998, but postponed it indefinitely in April 1998. The band plan for the GWCS band is illustrated in Exhibit 15-1 below.

Exhibit 15-1: GWCS Band Plan

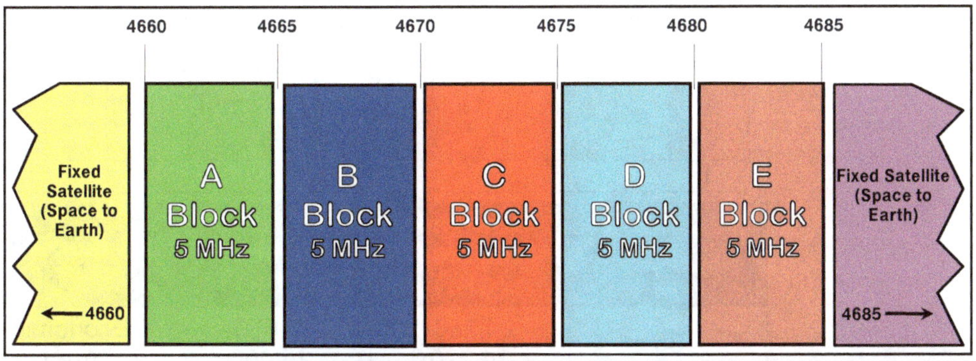

Block	Frequency	Amount	Type
A Bock	4660-4665	5 MHz	Unpaired
B Block	4665-4670	5 MHz	Unpaired
C Bock	4670-4675	5 MHz	Unpaired
D Block	4675-4680	5 MHz	Unpaired
E Bock	4680-4685	5 MHz	Unpaired

Source: FCC and Summit Ridge Group, LLC analysis.

2. Former Digital Electronic Message Service (DEMS): 24 GHz

This band was auctioned in the FCC Auction No. 56 in July 2004 for use with digital fixed service. Each license consists of a pair of 40 MHz channels (total 80 MHz). One license was issued in each of the 176 Economic Areas in the United States as defined by the Department of Commerce and four were issued for ten-year terms each for the FCC defined Economic Areas. A band plan for the 24 GHz spectrum is illustrated in Exhibit 15-2 below.

Exhibit 15-2: 24 GHz Band Plan

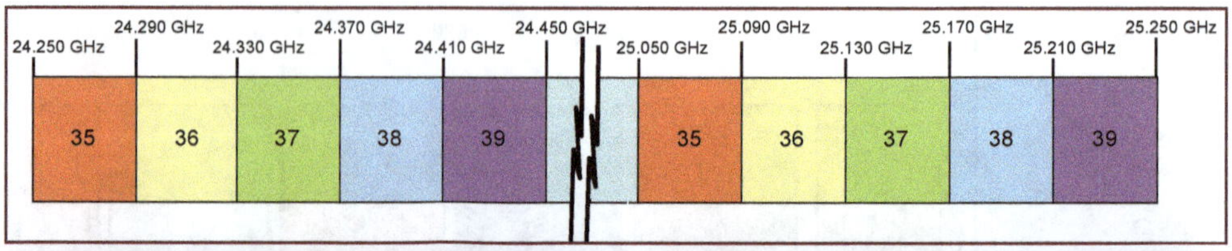

Channel	Downlink Frequency (GHz)	Uplink Frequency (GHz)	License Area	Licenses
35	24.250-24.290	25.050-25.090	Economic Area	176
36	24.290-24.330	25.090-25.130	Economic Area	176
37	24.330-24.370	25.130-25.170	Economic Area	176
38	24.370-24.410	25.170-25.210	Economic Area	176
39	24.410-24.450	25.210-25.250	Economic Area	176
Note: 80 MHz (paired 40 MHz segments) in each of five channels (Channel Numbers 35-39)				

Source: FCC and Summit Ridge Group, LLC analysis.

Local Multipoint Distribution Service (LMDS) spectrum was originally conceived to transmit digital television signals

3. LMDS: 27.5 to 30 GHz and 31.075 to 31.225 GHz

Local Multipoint Distribution Service (LMDS) spectrum was originally conceived to transmit digital television signals and for fixed wireless last mile delivery. Its range is typically limited to 1.5 miles (2.4 km) due to rain fade issues, i.e. strong rain dampens the signal's strength. However, with a larger antenna and unobstructed line of sight, the range can reach 5 miles (8 km). Although it showed great hope of providing effective "wireless cable" in the 1990s, this was never a commercial success. Most providers abandoned efforts to build wireless cable using this spectrum. Given the amount of available spectrum available in the LMDS bad, there is some industry discussion about using LMDS to deliver residential broadband.

LMDS spectrum was auctioned in FCC Auction No. 17 in 1998. The licenses were auctioned for each of the 491 Basic Trading Areas (BTA) into which the FCC has partitioned the country. The band plan for LMDS spectrum is illustrated in Exhibit 15-3 below. The A block licenses are 1,150 MHz while B block license are 150 MHz each. After some winning bidders withdrew their bids, part of the spectrum was re-auctioned in Auction 23 in 1999.

Exhibit 15-3: LMDS Spectrum Band Plan

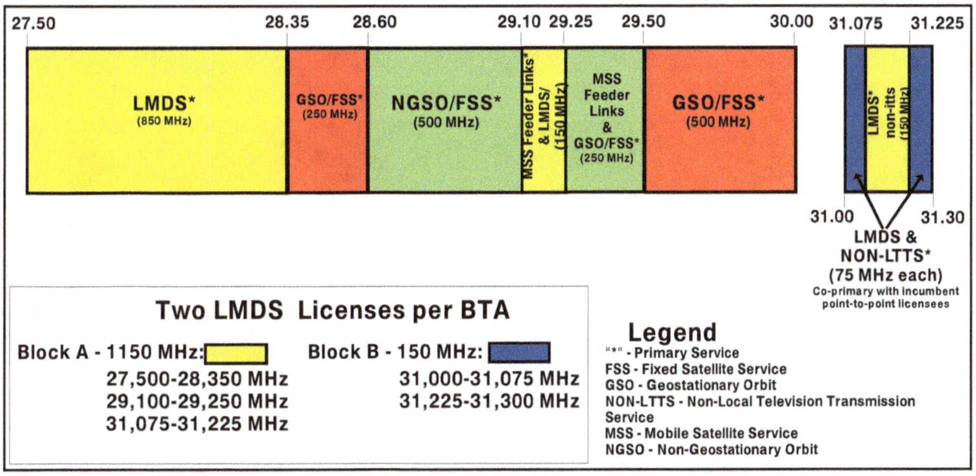

Block	Frequency	Bandwidth	Licenses
A Block	27.5-28.35, 29.1-29.250, 31.075-31.225	1,150 MHz	491 (BTA)
B Block	31.0-31.075 31.225-31.300	150 MHz	491 (BTA)

Source: FCC and Summit Ridge Group, LLC analysis.

The 39 GHz band is authorized for fixed point-to-point use for wireless backhaul and backbone communicat ion

B. Extremely High Frequencies (EHF): 30 GHz to 300 GHz

As mentioned in Section III.H, spectrum at this level and above is sparsely utilized. It is subject to severe rain fade and has a poor ability to penetrate buildings and foliage. Nevertheless, portions of the lower end of the range including the Q-band (33 to 50 GHz) and the V-band (50 to 75 GHz), which are allocated for satellite and short distance terrestrial broadband may be increasingly used in the future. An exception is the 39 GHz band described below.

1. 39 GHz band

The 39 GHz band, comprising frequencies from 38.6 GHz to 40 GHz was auctioned in FCC Auction 30 in 2000. The band is authorized for fixed point-to-point use for wireless backhaul and backbone communication. Licenses were issued in 172 Economic

Areas with fourteen 100 MHz (50 MHz paired) blocks in each area. The band plan for the 39 GHz band is illustrated in Exhibit 15-4 below.

Exhibit 15-4: 39 GHz Band Plan

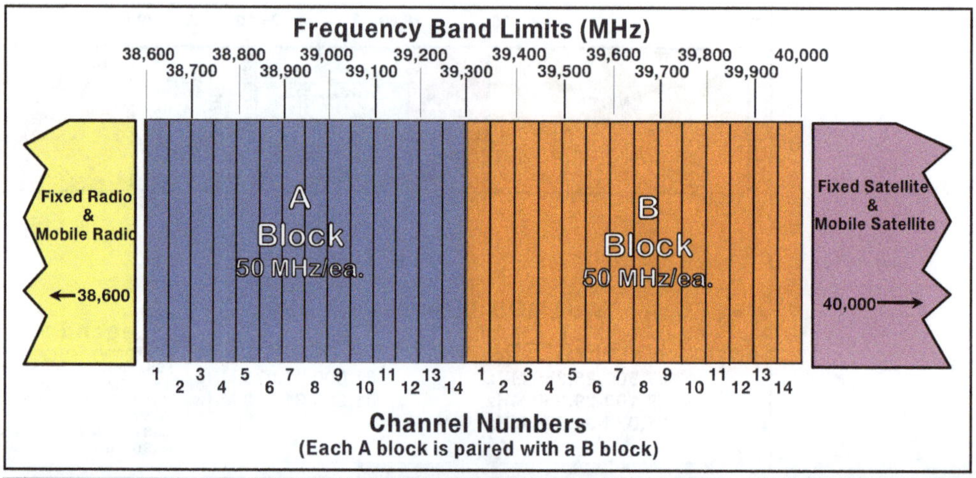

Block	Frequency	Bandwidth	Licenses
A Block	38.6-39.3 GHz	700 MHz	172 (EA)
B Block	39.3-40.0 GHz	700 MHz	172 (EA)

Source: FCC and Summit Ridge Group, LLC analysis.

APPENDICES

Appendix I: ITU Regions

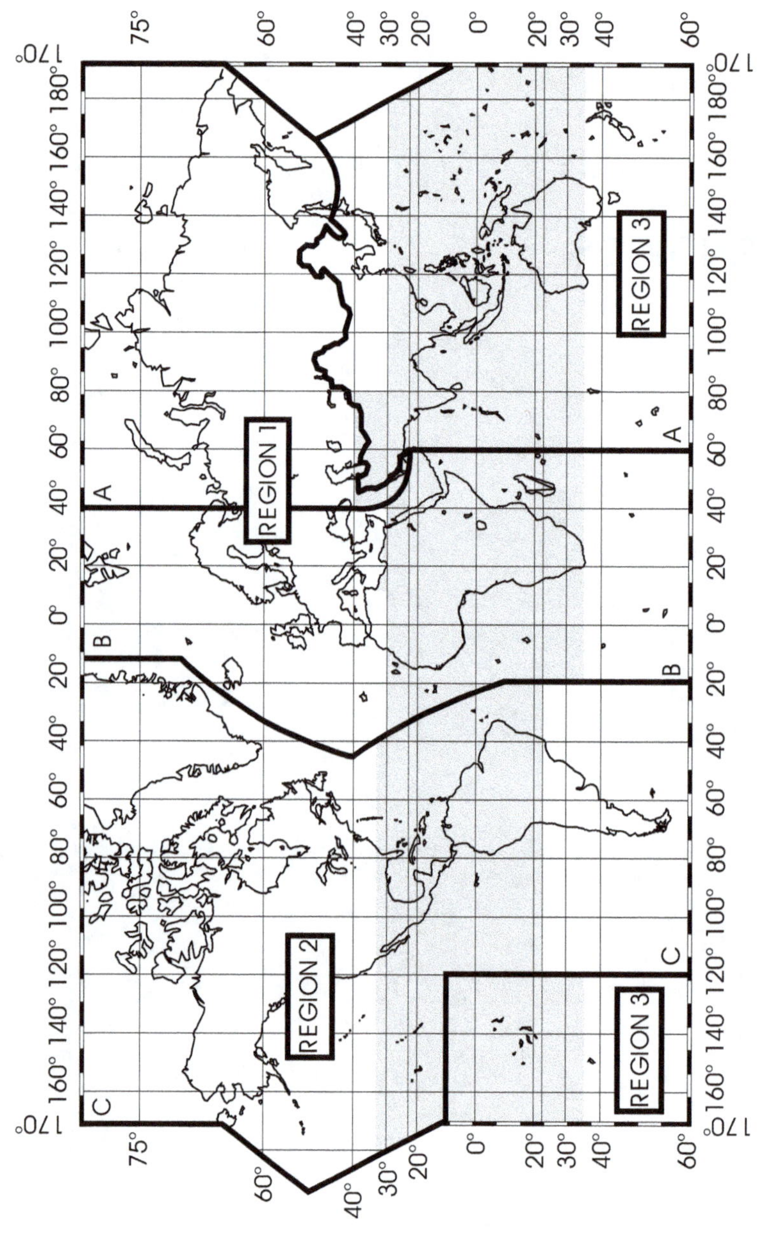

Figure 1: Map identifying Region 1, Region 2, and Region 3, as defined in paragraph 2.104(b), and the Tropical Zone (shaded area), as defined in paragraph 2.104(c)(4).

Source: ITU

Appendix II: Radio Spectrum Frequencies

Band name	Abbreviation	ITU Band	Frequency and Wavelength in Air	Examples of Uses
Tremendously low frequency	TLF		< 3 Hz > 100,000 km	Natural and man-made electromagnetic noise
Extremely low frequency	ELF		3–30 Hz 100,000 km – 10,000 km	Communication with submarines
Super low frequency	SLF		30–300 Hz 10,000 km – 1000 km	Communication with submarines
Ultra low frequency	ULF		300–3000 Hz 1000 km – 100 km	Submarine communication, Communication within mines
Very low frequency	VLF	4	3–30 kHz 100 km – 10 km	Navigation, time signals, submarine communication, wireless heart rate monitors, geophysics
Low frequency	LF	5	30–300 kHz 10 km – 1 km	Navigation, time signals, AM longwave broadcasting (Europe and parts of Asia), RFID, amateur radio
Medium frequency	MF	6	300–3000 kHz 1 km – 100 m	AM (medium-wave) broadcasts, amateur radio, avalanche beacons
High frequency	HF	7	3–30 MHz 100 m – 10 m	Shortwave broadcasts, citizens' band radio, amateur radio and over-the-horizon aviation communications, RFID, Over-the-horizon radar, Automatic Link Establishment (ALE) / Near Vertical Incidence Skywave (NVIS) radio communications, marine and mobile radio telephony
Very high frequency	VHF	8	30–300 MHz 10 m – 1 m	FM, television broadcasts and line-of-sight ground-to-aircraft and aircraft-to-aircraft communications. Land Mobile and Maritime Mobile communications, amateur radio, weather radio

Ultra high frequency	UHF	9	300–3000 MHz 1 m – 100 mm	Television broadcasts, microwave ovens, microwave devices/communications, radio astronomy, mobile phones, wireless LAN, Bluetooth, ZigBee, GPS and two-way radios such as Land Mobile, FRS and GMRS radios, amateur radio
Super high frequency	SHF	10	3–30 GHz 100 mm – 10 mm	Radio astronomy, microwave devices/communications, wireless LAN, most modern radars, communications satellites, satellite television broadcasting, DBS, amateur radio
Extremely high frequency	EHF	11	30–300 GHz 10 mm – 1 mm	Radio astronomy, high-frequency microwave radio relay, microwave remote sensing, amateur radio, directed-energy weapon, millimeter wave scanner
Terahertz or Tremendously high frequency	THz or THF	12	300–3,000 GHz 1 mm – 100 μm	Terahertz imaging – a potential replacement for X-rays in some medical applications, ultrafast molecular dynamics, condensed-matter physics, terahertz time-domain spectroscopy, terahertz computing/communications, sub-mm remote sensing, amateur radio

Source: Wikipedia and Summit Ridge Group, LLC analysis.

Appendix III: FCC Geographical License Areas

- **Cellular Market Areas (CMAs)** – 734 total areas. Used for:
  - Cellular Radiotelephone Service - 47 CFR § 22.909 (824 to 894 MHz: Blocks A-B)
  - Interactive Video and Data Service (IVDS) - 47 CFR § 95.803 (Segments: A (218.0 to 218.5 MHz) and B (218.5 to 219.0 MHz))
- **Basic Trading Areas (BTAs)** - 491 total areas. Used for:
  - Multipoint Distribution Service (MDS) - 47 CFR § 21.924 (2150 to 2162, 2596 to 2680 MHz: Channels 1, 2(2A), E1-E4, F1-F4, H1-H3)
  - Narrowband PCS - 47 CFR § 24.102 (901 to 902, 930 to 931 MHz: Channels 25 to 26)
  - Broadband PCS - 47 CFR § 24.202 (1850 to 1990 MHz: Blocks C-F)
  - Local Multipoint Distribution Service (LMDS) - 47 CFR § 101.1007 (27.5 to 31.3 GHz: Blocks A-B
- **Major Trading Areas (MTAs) 1990** - 51 total areas. Used for:
  - Narrowband PCS - 47 CFR § 24.102 (901 to 902, 930 to 931 MHz, 940 to 941 MHz: Channels 18 to 24)
  - Broadband PCS - 47 CFR § 24.202 (1850 to 1990 MHz: Blocks A-B)
  - Specialized Mobile Radio (SMR) 900 MHz - 47 CFR § 90.661 (896 to 901 MHz and 935 to 940 MHz: Channel Blocks A-T)
- **Regional PCS Areas (RPC) 1990**. Used for:
  - Narrowband PCS - 47 CFR § 24.102 (901 to 902, 930 to 931 MHz, 940 to 941 MHz: Channels 18 to 24)
  - Broadband PCS - 47 CFR § 24.202 (1850 to 1990 MHz: Blocks A-B)
  - Specialized Mobile Radio (SMR) 900 MHz - 47 CFR § 90.661 (896 to 901 MHz and 935 to 940 MHz: Channel Blocks A-T)

- **Economic Areas (BEA) 1990** – Aggregates MTAs into five areas. Used for:
    - General Wireless Communications Service (GWCS) - 47 CFR § 26.102 (4660 to 4685 MHz: Channel Blocks A-E)
    - Specialized Mobile Radio (SMR) 800 MHz - 47 CFR § 90.681 & 90.903 (806 to 821 MHz and 851 to 866 MHz: Spectrum Blocks A-C)
    - 220 MHz - 47 CFR § 90.761 (220 to 222 MHz: Channel Assignments A-E)
    - Location and Monitoring Service (LMS) - 47 CFR § 90.7 (3 Spectrum Blocks (1) 904.000 to 909.750 MHz and 927.750 to 928.000 MHz; (2) 919.750 to 921.750 MHz and 927.500 to 927.750 MHz; and (3) 921.750 to 927.250 MHz and 927.250 to 927.500 MHz)
    - 38.6-40.0 GHz Band (39 GHz) - 47 CFR § 101.64 (38.6 to 40.0 GHz: Channels 1-14 with paired blocks A-B)
- **Major Economic Areas (MEA) 1990**. Aggregation of BEAs into 52 regions. Used for:
    - Wireless Communications Service (WCS) - 47 CFR § 27.6 (Paired Channel Blocks A (2305 to 2310 MHz and 2350 to 2355 MHz) and B (2310 to 2315/2355 to 2360 MHz))
    - 929 MHz Paging - 47 CFR § 90.493 (Channels A-L: 929.0125 to 929.9625 MHz)
    - 931 MHz Paging - 47 CFR § 22.531 (Channels AA-BK: 931.0125 to 931.9875 MHz)
- **Regional Economic Areas (REA) 1990**. Aggregation of MEAs into 12 regions. Used for:
    - Wireless Communications Service (WCS) - 47 CFR § 27.6 (Unpaired Channel Blocks C (2315 to 2320 MHz) and D (2345 to 2350 MHz))
- **Economic Area Groupings (EAG) 1990**. Aggregation of BEAs into six regions not including the Gulf of Mexico. Used for:
    - 220 MHz - 47 CFR § 90.761 (220 to 222 MHz: Channel Assignments F-J)
    - 700 MHz - 47 CFR § 27.6 (746 to 764 and 776 to 794 MHz: Blocks A-D)

- **VHF Public Coast Station Areas (VCP) 1990**. 42 areas. Used for:
    - VHF Public Coast Stations 47 CFR § 80.371(c)(1)(B)
- **Television Market Areas (TMA) 1990**. 210 radio markets. Used for:
    - Determining areas for the cable must-carry/retransmission consent election. (47 CFR § 76.55)
        - Cable Television Service 47 CFR 76 Subpart D
    - Children's Television Programming Report (Form 398)
- **Emergency Alert Local Areas.** Used for:
    - Emergency Alert System 47 CFR § 11.21
- **Standard Metropolitan Statistical Areas (1980) SMSA1980**. SMSAs are unique in that they are not continuous across the nation. Also, they aggregate minor civil divisions (towns and cites) as opposed to county-equivalents in New England. They consist of 323 areas. Used for:
    - Original* Digital Electronic Message Service (DEMS) - 18 GHz and 24 GHz
- **Nationwide Area (NWA) 1990**. Used for:
    - Narrowband PCS - 47 CFR § 24.102 (901 to 902 MHz, 930 to 931 MHz and 940 to 941 MHz: Channels 1 to 11)

Appendix IV: Glossary

- **3G:** Third Generation of mobile technologies. 3G decreased spectrum costs, enabled customers to achieve higher data transmission rates (mobile systems: more than 384 kpbs; stationary systems: more than 2 Mbps) and improved efficiency in the use of spectrum. Comes in different transmission formats such as W-CDMA, UMTS, CDMA 2000, TD-CDMA, DECT and Mobile WiMAX.
- 3GPP: (Third Generation Partnership Project). An international umbrella group for telecom regulation coordination.
- **4G:** Fourth Generation of mobile technologies. 4G decreased spectrum costs even beyond 3G. It enables customers to achieve higher data transmission rates – as high as 1GB/Sec.
- **AMPS:** Advanced Mobile Phone Service.
- **ATC:** Ancillary Terrestrial Component. Using satellite telephony spectrum to let MSS provide service terrestrially at the same time as using it to provide service via satellite. The use of ATCs in Mobile Satellite Services was introduced in 2003.
- **AWS:** Advanced Wireless Spectrum: 1710 to 1755 MHZ for uplink and 2110 to 2155 for downlink. In the United States this was broken-up into six blocks (A-F) that have been subject to auctions since 2006.
- **Bandplan:** A regulatory plan for using frequencies in a certain frequency band. It includes numerous technical details about the band including the channel division, licensing, modulation and other details.
- **BRS:** Broadband Radio Service: Used for general-purpose networking. Uses microwave frequencies between 2.5 and 2.7 GHz. Increasingly in demand for mobile wireless applications.
- **C-band:** 3.4 to 4.3 GHz (receive) and 4.25 to 6.425 GHz (transmit). Used primarily for satellite transmission and desirable in tropical areas due to its resistance to rain fade. Many countries are considering reallocating the lower portion of the receive band for terrestrial broadband.
- **CEPT:** European Conference of Postal and Telecommunications Administrations. A European regional coordinating body for European telecom and postal issues.

- **CDMA:** (Code Division Multiple Access). A spread-spectrum technology that allows multiple users to transmit over the same physical frequency channel, but prevents interference by coding each transmission.
- **CEPT:** (European Conference of Postal and Telecommunications Administration). European regional telecommunication regulating body.
- **CITL:** (Inter-American Telecommunication Commission). A regional telecom coordinating body for the Americas.
- **Communications Act of 1934:** Authorized the creation of the FCC.
- **Duplex Gap:** The space between the uplink and downlink frequencies in paired spectrum.
- **EBS:** Educational Broadcasting System.
- **EDGE:** (Enhanced Data Rates for GSM Evolution). An advanced version of GPRS.
- **Downlink:** Transmission from the base station to the user.
- **FCC:** The Federal Communications Commission. An independent agency of the U.S. government that works in the areas of broadband, competition, the spectrum, the media, public safety and homeland security.
- **FDD:** Frequency Division Duplexing. A technology that uses different frequencies for the send and receive signals to minimize interference. FDD is the most common duplexing technology in the United States. This is in contract with TDD (Time Division Duplexing).
- **FSS:** Fixed Satellite Services.
- **GPS:** Global Positioning Satellite. A satellite systems run by the U.S. defense department. It emits a timing signal that can be triangulated to determine geographic position.
- **GPRS:** General Packet Radio Service. GSM (see above) expanded for data.
- **GSM:** Global System for Mobile Communication. A standard developed by the European Telecommunications Standards Institute for protocols for second generation (2G) mobile phones.
- **GWCS:** General Wireless Communication Service. A communications band at 4.660 to 4.685 GHz.
- **Hertz:** one hertz is one frequency wave. Frequency is frequently referred to by the number of waves (hertz) produced in a second.
- **IEEE:** Institute of Electrical and Electronic Engineers.
- **ITU:** (International Telecommunications Union). Global coordinating body for

radio spectrum. The ITU is based in Geneva, Switzerland.

- **ITU-R:** The division of the ITU that handles most spectrum issues.
- **Ka-band:** 26.5 to 40 GHz. This satellite band has historically suffered from significant weather related interference. However, new higher-powered satellites over the past 10 to 15 years have largely overcome this problem. This band is heavily used in state of the art satellite broadband systems.
- **Ku-band:** 12 to 18 GHz. Used primarily in satellite communications. Most commercial satellite television signals in the United States and Europe are transmitted in Ku-band. Although not as resistant to rain fade as C-band, its higher frequency allows consumers to use much smaller reception antennas.
- **L-band:** 1452.96 to 1492.624 MHz. Used for satellite applications including satellite radio and MSS applications.
- **LMDS:** Local Multipoint Distribution Service. Used for broadband wireless. Originally intended for wireless digital television. Uses the 31.0 to 31.3 GHz frequency. Now being considered for broadband backhaul and last mile broadband access in certain areas.
- **LTE:** Long Term Evolution. A fourth generation version of GSM with data rates as high as 1 GB/Sec.
- **LORAN:** (LOng RAnge Navigation). A low frequency timing and positioning system that has largely been replaced by global positioning satellites (GPS).
- **MMDS:** Multichannel Multipoint Distribution Service (see BRS).
- **MSS:** (Mobile Satellite Services). For provision of mobile services using satellites. Only since the introduction of ATCs in Mobile Wireless Services in 2003 has there been an integration of both satellite and terrestrial uses in one frequency band. Bands that provide MSS are the S-band and L-band.
- **Narrowbanding:** A regulatory reduction in the amount of spectrum allocated to a frequency band.
- **NTIA:** (National Telecommunications and Information Administration). An agency within the department of commerce that regulates and administers spectrum allocated for government use.
- **Ofcom:** The British telecom regulatory body.
- **Orbital Slot:** The location in space allocated for a satellite in a specific frequency band.
- **Paired Spectrum:** Spectrum that is allocated in two blocks, one authorized for uplink transmission and one authorized for downlink transmission.
- **PCAST:** President's Council of Advisors on Science and Technology. In

2012, this group issued an influential report recommending spectrum sharing as a means to increase spectral efficiency.

- **PCS:** (Personal Communication Services). In North America, wireless voice and data services on the 1850 to 1990 MHz band.
- **RF:** (Radio Frequency). Typically refers to spectrum between 3 kHz and 300 GHz.
- **S-band:** 2 to 4 GHz. Used for satellite MSS applications.
- **Smart Radio:** A radio that can use multiple frequencies and determine the appropriate frequency to use based on the design criteria.
- **TDD:** (Time Division Duplexing). A technology that separates the send and receive signals by time. This technology is advantageous when usage is asymmetrical. TDD is common in China.
- **TDMA:** (Time Division Multiple Access). A channel access method that allows users to share the same frequency by dividing transmissions into different time slots.
- **UMTS:** (Universal Mobile Telecommunications System). A 3G version of GMS (see above).
- **Uplink:** transmission to the base station.
- **VSAT:** (Very Small Aperture Terminal). A small satellite dish used for a variety of satellite communications needs including corporate networks.
- **Wi-Fi:** A format for transmitting broadband data across short distances to unlicensed consumer devices.
- **WCS:** Wireless Communication Services.
- **White Spaces:** Unused frequency in the broadcasting bandplan.
- **WiMAX:** Worldwide Interoperability for Microwave Access. A format/standard for information transmission, mainly to provide fixed wireless broadband services. WiMAX can be used in a variety of spectrum bands, such as 700 MHz, 2.4 GHz, 3.5 GHz and 5.8 GHz, although the main ones for which there is support are the WCS and BRS bands.
- **WRC:** (World Radio Conference). A event run by the ITU that revises and reviews Radio Regulations, the international treaty governing the use of radio spectrum and satellite orbits.